Paul Robinson

THE OUTER LIMITS OF LIFE

John Medina

A Division of Thomas Nelson Publishers
Nashville

To my mother
for giving me the keys
to so many doors
and
to my wife
for giving me the courage
to walk through them

Published in Nashville, Tennessee, by Oliver-Nelson Books, a division of Thomas Nelson, Inc., Publishers, and distributed in Canada by Lawson Falle, Ltd., Cambridge, Ontario.

Library of Congress Cataloging-in-Publication Data

Medina, John, 1956–
The outer limits of life : a molecular biologist looks at life and the implications of genetic research / John Medina.
p. cm.
ISBN 0-8407-9114-3 (hardcover)
1. Life. 2. Body, Human—Religious aspects—Christianity. 3. Body, Human. 4. Life (Biology) 5. Biology—Religious aspects—Christianity. 6. Genetics—Research. 7. Bioethics. 8. Christian ethics. I. Title.
BT696.M435 1991
113′.8—dc20 91-16702
CIP

Printed in the United States of America.

1 2 3 4 5 6 — 96 95 94 93 92 91

Contents

Foreword

John Medina has, I believe, written a most important book.

Why is it important? First, because his subject is vital. Life issues—abortion, euthanasia, and now the harvesting of fetal tissue and organs—have earthquake status in today's Western world, for here the tectonic plates of historic Christianity and post-Christian secularity grind against each other in direct and traumatic collision. Dr. Medina is a molecular biologist, which means that everything involved in cell research—biochemistry, neurology, genetics—has become his business, and he can bring them together in an authoritative and comprehensive way. *The Outer Limits of Life* is a systematic, thorough briefing on the relevant scientific facts, and as such is invaluable.

Second, Medina's standpoint is honest. When he is sure of something, when he is not sure of something, and when he feels torn between his scientific responsibility and his Christian calling, he says so. His book is not propaganda (there is plenty of that already, on both sides); rather, it is a reflective report from the frontier, where cell research has been going in at a great rate for the past two decades and seems likely to continue just as fast. The integrity of the entire discussion is outstanding.

Third, the professor's style is delightful. He has a great gift for nontechnical exposition: his reader-friendly way of simplifying things has resulted in a book, on themes usually cloaked in dark jargon, which any reader from nineteen to ninety can follow and digest. Even I, a longtime refugee from science, who thankfully turned his back on chemistry, physics, and biology at age thirteen to study Greek, was able to read it with pleasure and understand it without pain; and if I could do that, so can you.

Fourth, John Medina's stance is robustly and clearheadedly Christian. He does not offer instant answers to all questions in the life-issues debate (that, indeed is one aspect of his wisdom), but he has a Bible-based, well-digested way of facing up to them all, laying out the facts one must know about them, and squelching shoddy and unworthy arguments about them. His attitudes are a model of how a scientist who is also a Christian should think his way through things.

So I gratefully salute a very fine and fruitful piece of work.

J. I. Packer

Chapter 1

A Matter of Life and Breadth

Tumors and Other Strangers

One can only imagine the horror on the face of the first person who discovered the following fact:

You can surgically remove certain tumors from the sides of people, dissect those tumors, and find fully formed fingers deep inside them. And teeth. And bits of bone. You can find cells that make up the lens of the human eye, skin that has real hair on it, and cells that are trying to beat like a heart. You can find virtually the entire range of human tissues represented in some form with this kind of cancer. Physically, the tumor looks like a biological tomb whose corpse has been disturbed by grave robbers and wild animals. These cancers, called *teratoma tumors,* are the source of a great deal of excitement to many research scientists.

That's right. Excitement.

From a distance, this mixed-up bag of cells looks like it was trying to become a human being. But something went horribly wrong with the effort, and the whole thing collapsed into a disorganized mass of human parts. Sort of like cancer imitating life. Or perhaps imitating Mary Shelley. Different authors, mostly writing for tabloids, have used the biology of this tumor to report the progress genetic engineers are making in building a human being from scratch.

The tumor isn't trying to imitate anything, of course. In this instance, certain human cells are responding to specific molecular cues and, as a reaction, form eerily familiar structures. These are predictable, solidly biochemical processes full of the magic—and the reality—of any good physical phenomenon.

From a research point of view, ignoring the window the teratoma tumor has opened to us is impossible. Why is that? One reason is that this anomaly allows us to glimpse into processes considered by many to be the Holy Grail of developmental biology; we can watch how garden-variety cells turn into specialized tissues that eventually make such genetic masterpieces as eyes and hearts and brains. Another reason is a social one. Tucked inside this innocent investigation of cellular processes is a bombshell of a question: What mechanisms turn human tissues into human beings? The answer has implications that reverberate far beyond the laboratory bench. That's why the examination of the teratoma is so exciting. This tumor contains a rich source of genetic recipes that, in part, may tell us how this transition occurs.

As a practicing research scientist, however, I feel the excitement about this tumor end as soon as I leave the laboratory. It is frightening to wonder out loud how human cells organize their interior molecules to carve out human individuality. When we look for the differences between human lives and human tissues, we are literally talking about life-and-death issues. Such issues, applied to certain genetic processes, risk being melted into piles of scientifically unanswerable questions. And that's the scary part. I love solid black-and-white answers, especially when they are gestated in solid black-and-white questions. But when such questions give birth to answers of a different color, how we should perceive them becomes much less obvious to me.

I remember having certain feelings about this issue during the first funeral I attended. When I was in the third grade, my teacher became very ill, and she died in the middle of the school year. I remember the event very clearly, primarily because of my best friend. Actually, because of my best friend's mother. She had just had a baby. And my best friend and his mother and the baby came to the funeral and sat next to me and my mother. The baby did nothing but scream his little lungs out. Even when the service was ending and we filed by the open casket to pay our respects, the baby was still wriggling and making loud noises.

Years later, I thought about the juxtaposition: A dead, lifeless person was sharing the same room with a lively, exuberant little bundle of humanity. What was the difference between the two states? Why would roughly the same genetics be so conspicuously silent in one place and so conspicuously robust in another? What are the differences between a dead teacher in a casket and a noisy infant in a mother's arms. And what do both of these have to do with a nightmarish "semihuman" in a tumor?

These questions force us to confront a single, and rather ancient, identity crisis: discerning the sentience in human life. These days, we have many tools at our disposal with which to extract insights, including some rather powerful genetic ones. And perhaps that is fortunate. Whether we go to funerals or baby showers, most of us try to get some perspective on what these differences mean in our lives. And the purpose of the book in your hands is to find out exactly what they are.

On the Matter of Loving Your Pet Cat

I was once seated at a restaurant next to a very famous genetic investigator who had won lots of awards for his research. Which was amazing. Not because he had won lots of awards; the scientist's brilliance bordered on the incandescent. It was mostly amazing that I got the chance to sit next to him. In our two hours together, he told me his thoughts about biological life, from subtle genetic processes to conspicuous ecological concerns. The discussion was one of the most fascinating I have ever had.

But we didn't stick to the latest cloning techniques or the destruction of the Amazon rain forest. We also talked about personal issues. He told me that before he was married, he owned a cat. He was quite fond of the animal, and when it died, he grieved for it as he might any lost friend. Even though he is now married and has a daughter, he still grieves over his cat, and the intensity of the emotion surprises him to this day. He wondered out loud how he would feel if he lost his daughter or his wife. And then he began to wonder why he should feel anything at all. The conversation drifted to the topic of worth and the reasons people so close to us mean so much.

I asked him why he valued his wife and daughter and the memory of his cat. In an attempt to understand his point of view, I posed to him a hypothetical circumstance. I asked him to imagine that his cat was still alive and that both of them were in Cambodia around the time of the Khmer Rouge dictatorship. Suppose a Khmer Rouge guard came bursting through the door of his house with an AK-47 assault rifle. The guard was dragging a little boy about nine years old and had the gun aimed at his head. The guard shouted in broken English that either the cat had to be killed or the nine-year-old boy had to be killed and that the scientist, as an American, must decide between the two. Which one would he choose?

My dinner companion put down his fork, looked out the window, and responded with an answer that made my blood run cold. He said, "That's easy. I would choose to let my cat live. I would let the boy die."

He then told me why. He had a relationship with the cat but no relationship with the boy. He saw no intrinsic worth in the human because of the more complex physiology. Both were collections of genes and, in his mind, equally worthy biochemistries. The value judgment was made primarily on the basis of familiarity, on a personal relationship with one particular group of genes, not on any loyalty to the species. We ended our dinner, and our conversation, on the idea that his universe was given value simply by what he thought of it. And that when he died, those values, for all intents and purposes, would die with him.

I went away from that conversation with a very heavy heart. I also went away trying to think about great social forces. I wondered if we, as a body of people, needed to talk about the presence of human life in its most intimate and exposed form. That is, we needed to talk about ourselves in terms of our genes and how research into those genes might change the nature of view of ourselves and our neighbors.

I went back to the laboratory and told my colleagues about our conversation. Most of them disagreed with the famous scientist. They asserted that while none of us needed to place humanity on an altar to render it inaccessible to inquiry, we need not place it in the gutter to render it understandable, either. Or trivial. But that a man of such stature could have such a thought also produced a great deal of controversy for some in the group. And in the midst of an already difficult issue, a great deal of confusion.

This chapter functions as an introduction to this difficult idea of human life and human death, and perhaps describes a map for the intellectual layout of this book. In an effort to more narrowly define our topic, I have organized certain ideas around simple questions. The discussion can be distilled to four of them: (1) What is biological life? (2) What is biological death? (3) What is human life? (4) What is human death? I want to talk about the physical mechanisms that must at some level harbor human individuality. Even though you are reading this from a scientist's perspective, it is not meant to address the sometimes inaccessible scientific community. Rather, it is an attempt to address something we all experience the instant we wake up and feel the sand in our eyes. It is an attempt to talk about life and all that this gift reserves for us and for those we hold dear.

From the outset I realize that not all of life can be described in biochemical terms. However, our human experience depends on this biology like a family vacation depends on a station wagon. So, for our purposes, a biological description of what makes bugs crawl, fish swim, and humans eat potato salad makes a good starting point.

That means we lose the battle for narrowing down our topic. Nonetheless, we have to understand quite a bit about the chemical engines that cause our biology to plow through our environment. Ultimately, we will talk about genes and chromosomes, the biochemical air traffic controllers residing in most organisms. We will consider how their constituent parts contribute to the functioning and passage of biological information. The idea is to see if such a description can fully explain how organisms acquire the characteristics of life. Or see if life survives as a concept after such mechanisms have been described.

That explanation wouldn't answer our question if we stopped there, though. To appreciate whatever life turns out to be, we must contrast it with its absence; that is, we must discuss biological death and talk about it in the light of the same mechanisms used to describe life. If it is possible, we will use the contrast to establish a framework on which to build an initial definition of biological existence. This is always convenient. Establishing a framework is scientists' way of beating around the bush with questions we haven't the slightest idea how to answer. However, such nebulosity may reflect the task more than the effort applied to it. At this point in our development, scientific or otherwise, running in circles may be all we are capable of doing.

This huffing and puffing around the intellectual racetrack will, I hope, give us strength to talk about ourselves. Our next task will be to superimpose the quality of humanness onto our definition of life. To accomplish this objective, we will have to understand the mechanisms of its headwaters: the wriggling, thriving miracle of human conception. And then we will have to describe the genetic fruit of that conception; we will have to outline how a two-celled collision can give rise to a crying, pooping, wonderful little baby. This examination will allow us to consider more carefully the role of unfolding complexity in the establishment of existence and perhaps give us a hint about when humanity, if it was ever absent, finally arrives.

Having looked at the biology of human presence, we are then in a position to contrast it with its absence, also. Specifically, we must understand when the quality of human sentience no longer exists as a quality of biological life. We have to talk about something so personal that probably not even Geraldo Rivera will discuss it willingly (at least if he doesn't want any lawsuits). That is, we must talk about the biology of our thought life. In most ways, the seat of our humanity begins and ends with what comes out of our brains. Since we have defined the absence of that humanity as a cessation of such a flow of information, we must discuss what that means in solid biochemical terms and determine its impact, if any, on discerning human tissue from human life.

Most of this discussion will be marinated in my viewpoints and prejudices. That is perhaps an unavoidable consequence of addressing issues that become so personal so very quickly. From my perspective, to be complete, we will have to conclude with a solidly sovereign discussion about an even more solidly sovereign God. I realize that one strolls across risky ground by putting an island of belief in a sea of cold hard data, even if it is a belief in frigidness—which it is not. I also realize that addressing this topic pops me right out of my scientific expertise; I am equipped professionally to deal with genes, but I am not equipped professionally to deal with their religious connections. Theology is not my master, even if Jesus is my Lord. Nor are exegetical dissertations on earth-demolishing ideas. I don't even like big words. My only hope is to communicate how I couple my research with how I perceive my God and pray that the description will be clear, unambiguous, and placed on more solid footing than human love for a pet animal.

Better Clones and Gardens

Discussing general definitions of biological life and death and then narrowing those definitions to specific human life and death can be confusing. Before we begin, I would like to talk about this confusion and to discuss a brand new technology, one that promises to add much gray to a landscape many people would be more comfortable painting black and white.

Despite all our sophistication, most of us are still in the dark regarding the great engines of human life and human death. It is easy to understand this ignorance. For all the centuries of experience we've had to observe and discuss The Beginning and The Ending, we have no firsthand information about either event. For example, no one alive who has experienced conception personally can remember any part of the event. Since two joined cells have yet to form a brain, let alone an opinion, such amnesia is really a small wonder. In addition, each person who knows about permanent death firsthand is also still quite permanently dead. This vacuum leaves us, who aren't experiencing either event, wishing to discuss both. And without any rational basis for doing so.

That's all well and good. Rationality has never been a prerequisite for possessing a point of view. That's why there are as many opinions about this topic as there are people on this planet. If those opinions are stoutly held, there tends to be lots and lots of conflict since people who know they are right make a majority of one. This only exposes how easily we pour extremely heavy questions onto extremely fragile supports. Or on supports whose existence can be extremely difficult to prove.

It would be nice if a physical tool existed that would help all 5 billion opinions find a common watering hole. Looking for such a device certainly is tempting. We have had a tendency to use physical tools to solve problems ever since we learned it was easier to bonk an antelope over the head with a stone than it was to outrun it. In the twentieth century, we have been inventing new variations on old tools and changing the nature of what a tool is. And that has produced a series of technologies that had, as little as a decade before, been the exclusive realm of science fiction.

Some of these technologies have been produced in the field of biology. They have come under the general category of genetic engineering.

Or scientific earthquake, for short.

Genetic engineering or, more properly, molecular biology has taken the world of research biology and turned it on its ear. Cracking the genetic code has given us such intimate access to organisms that parts of them can be artificially reconstructed and replaced into the organism like false teeth. Or a clutch plate. Cutting and pasting genes as if they were tagboard have allowed us to reconstruct creatures that used to exist only in the minds of Goya and Hieronymus Bosch. These techniques have increased the availability of numerous medicines and dramatically reduced their cost. They have given hope to thousands of people previously sentenced to a terrifying spot on genetic death row and given even more thousands the ability to know if they were there in the first place. But it has done much more than give us an accurate prediction of future health.

Genetic engineering has yielded to us the secrets of understanding why we pass on certain traits to our children. And as a result, we have the potential to genetically resculpture ourselves and our children. It is technology that, if treated with any of the moral insight we treated the splitting of the atom to, should send all of us running for the Pepto-Bismol.

This powerful new set of tools has also pushed certain lines of research to the precipice of some people's moral convictions. And many are left wondering if science, which can define so clearly the edges of a problem, might define as clearly the edges of a solution. But is this science the right tool for answering questions about human sentience? Can modern-day biological expertise get us closer to understanding the difference between the arms at our sides and the people to which they are attached? Can science describe a process so morally compelling that all 5 billion people throw up their hands at once and say, "Aha, that is the difference between human beings and human cells"? We are used to having science answer many of our challenges and solve many of our problems. Is it capable of answering the ultimate identity crisis?

Answering that question may give us insight into our own confusion regarding human biological identity. One of the great characteristics of the twentieth century is the incredible social elevation we have awarded technology and the natural sciences. This elevation has become so powerful that we have mistakenly believed that technology and science could participate in border skirmishes with certain moral questions. To see why this is an error, we need to know something of science's strengths and a great deal about its limitations. We will talk as much about the nature of science as about the substance of what it presents.

You may already be extremely familiar with the topic. If you have had children, you are the ultimate genetic engineer. I am not sure background makes that much difference, however. Although most of us have not undergone the experience of earning a doctorate in molecular biology, philosophy, or theology, we all have undergone the experience of being human on planet earth. Cold, objective, and narrow-minded scientists laugh and love and get married and have children just like anyone else; even our ability to value cats appears at times to interfere with our ability to be human. Besides, even the slowest foot is elevated to some level if it enters a jogging contest. In the end, it doesn't matter who you are; your assignment to the human condition automatically qualifies you to assist in the investigation of this question. I no longer believe these issues can be decided arbitrarily by technologies we can barely understand anyway. Rather, all of us need to appreciate the nature of God's great investment in human biology. Even if it means holding our intellectual breath and taking a plunge into the most personal aspects of who we are.

Whatever that is.

Chapter 2

Have You Driven a Ford Lately?

Still Life

The questions concerning the nature of life and death have been so overworked that the discussion almost needs union representation. A superficial examination reveals only the obvious: a dead sea gull on the beach unmistakably lacks something compared to her colleagues flying noisily above. With all our technology, we seem no closer to answering these questions than when we were half-naked pharaohs building pyramids. And it remains frustrating because this dead sea gull's fate is so easily and so inexorably available to all of us. What are the dramatic differences between these two states? What are the nature of the forces driving the organisms that populate this planet? How do those differences intersect with human biology?

This chapter attempts to rework this discussion from a genetic engineer's point of view. Since Western medicine typically tears down a creature to build up an idea, we will discuss human life by examining the contributions of individual organs and cells. But we will mostly focus on the genes and chromosomes, those lilliputian biochemicals that are in reality the ringmasters of human biology. And I would like to start with an unlikely source for biological understanding, the typical American automobile.

I had a friend named Larry who, in a patriotic moment, decided to purchase a car made exclusively in the United States. Larry shopped around, selected a star-spangled, all-American model, and bought it. While waiting to pick up his new vehicle at the dealership, he glanced through a car magazine. The magazine contained an article that described his newly purchased vehicle, including where the various internal parts of his model were made.

To his horror, Larry found that his car engine was not made in the United States but was molded in Korea and assembled in Japan. He discovered that his carburetor was manufactured in a factory in Germany and the car radio came from Taiwan. The metals used to make the body of the car were originally mined somewhere else, molded somewhere else, and then shipped to the United States. Worse, he read that certain cars he called foreign had more parts originally manufactured and assembled here in the United States than anywhere else.

Larry slowly began to understand that his original question should not have been, Is this car made in the United States? It should have been, How much of this car is manufactured and assembled in a foreign country before coming to the United States?—a more specific and quite different question. He didn't know it at the time, but he ran head-on into a subtle and amazing principle of science that day: attempting to find the answers to a particular question often changes the nature of the question.

To comprehend the nature of biological life, many people believe it is important to understand the characteristics of the organs, cells, and molecules of which it is composed. That strategy is adequate but has the same problem as Larry's approach to buying an American car. Trying to discern the answers to the substance of biological life changes the nature of the question. The very constituents of the organisms themselves confuse the issue. Most processes that control biological existence are extremely complex, are not particularly well understood, and possess dumbfounding capabilities.

For example, you have at your disposal over 250 million individual skeletal muscle fibers, divided into roughly 430 groups, that help you get out of bed every morning. This collection is so powerful that if all the cells were rigged to pull together in one direction, the average human could lift almost twenty-five tons. To help you eat your breakfast, you have cells in your stomach that make an acid so powerful that if it got out of your mouth, it would burn a hole in your rug. To help you read your paper in a hurry, you possess a brain that can send a signal that crosses the length of a six-foot human 177,000 times in one second.

These organs and accompanying fluids are the substances that make daily human life possible. In an attempt to identify life, we must determine what engine drives these organs and coordinates the cells of which they consist. And to start this investigation, we have to look under the hood of a typical adult human being.

Power in the Blood

One great characteristic of human biological existence is its stubborn staying power in the face of very narrow, almost finicky, physical requirements. For example, by mid-afternoon of a typical day, you've pumped about 1,000 gallons of blood through your heart. That's about 55 million gallons in an average lifetime. In the time it took to pump those 1,000 gallons, it also beat 52,000 times. And if you did any heavy exercise today, you could have caused this muscle to beat almost ten thousand times an hour. Yet for all its might, the human heart is quite small. It is about the size of your fist and weighs less than a pound. This little organ never sleeps and never rests. If you take care of it, this muscle will continue to beat for better than seven decades—more than 2.5 billion times.

This take-it-all midget of a pump must squirt blood through your body. A liquid that is paradoxically anything but sturdy, blood must have only a certain number of salt molecules, it cannot be too thick or too thin, it cannot contain too much of a single gas, and it will not tolerate the presence of foreign organisms. Like hair shampoo, it has to be within a certain pH range. A pH range is nothing more than a measure of the concentration of a certain chemical (really a hydrogen ion) in a substance.

To understand how pH can affect your biochemistry, consider what happens to your mouth when you eat a sour lemon peel. After the first bite, the pH in your mouth goes down from a normal pH of about 7.0 to somewhere around 4.0, a change most interpret as painfully sour. If there is too much free hydrogen ion—and lemon peels have way too much of it—your oral pH will become acidic, and the inside of your mouth will feel like crinkled cellophane. That may not sound life-threatening to you, but if that sudden pH drop occurred in your blood instead of your mouth, you would die a very miserable death. Your blood must stay in a pH range between 7.35 and 7.45 for the entire span of your life. Because of such pickiness, your life is precariously balanced on the ability of your blood to regulate something as seemingly obscure as its hydrogen ion concentration.

But blood requires even more aristocratic accommodations to function. Like most human tissues, it must stay within a very narrow temperature range. The body has developed elaborate mechanisms in circulatory vessels to keep the blood and its surrounding tissues warm and happy. If you get too cold, these vessels constrict, decreasing the flow of blood to the skin, preserving precious heat within. If you get too warm, the vessels dilate and the blood rises to the surface to dissipate the heat. That's why people often turn red when they are too hot. This requirement for internal heat regulation is shared by most mammals and is the source of the term *warmblooded.* None of us can stand to live in a climate unprotected if the temperature shifts by a few tens of degrees. Whether one examines salt molecules, pH, or temperature, blood exhibits some fairly fragile physiological requirements. Yet it is pumped by an organ whose physiological athletic feats are anything but fragile.

Pink Floyd Was Wrong About Thought Control

Numerous organs besides the heart exhibit great endurance and great specificity. Most of these systems continually interact with each other. Some of the interactions are cooperative, such as your throat's swallow reflex, which coaxes your morning omelet past your teeth into your stomach. Some exist as a monitoring system. White blood cells, for example, make the bloodstream suffer through round-the-clock inspections looking for foreign invaders. These cells have the ability to sound a quick and relatively massive immunological alarm if they detect intruders.

Perhaps the ultimate monitoring system of the human body resides in the brain and its attendant nerve structures. Close your eyes for a minute and extend your right hand. Without wiggling, you can tell exactly where your fingers are. You can also assess how far your hand is from your chest and the angle of your arm in relation to the floor. Now how do you know that information? You cannot hear your hand. You had your eyes closed so you could not see anything. You could have blocked out most external stimuli, and you would have been able to tell where your hand was.

The reason for this perception is simple. Your entire body has been bugged, just like a foreign embassy under surveillance by a hostile country. Your brain has special sensors hard-wired into your muscles. They continually report to your head day and night where every part of your body is in relation to itself and your environment. Your toe is continually relaying information about its location relative to the floor beneath it. Your arm is incessantly transmitting information about its position, angle, and distance away from the rest of the body.

Scientists have given this phenomenon two names, *proprioception* and *kinesthesis*. This information gives the brain an overall three-dimensional reference about your body's position in its environment. The brain taps into this information whenever a new movement must be performed; there is evidence that it keeps a permanent memory of nerve firing patterns for the various limbs. This memory may help to explain the experiences of amputees who feel sensation or pain in limbs that no longer exist. The pattern of the limb's former location may be remembered by the local nerves and transmitted to the brain.

Though the brain exerts a great deal of control over the body, its dictatorship is not all-encompassing. Take, for example, the interaction between the heart and the head. The brain can regulate the heart by slowing it down or speeding it up. It can do everything but jump start the heart after it has stopped beating or prevent beating once it commences. The reason for this limitation is that the heart has a "brain" all its own. Special nerve centers, collectively termed the *conduction system,* are sprinkled throughout this continually beating muscle like dandelions on a lawn. The heart displays a pumping action in response to this system even if no external command tells it to do so. A heart can be removed from the body, and it will continue to beat. In the laboratory, we have been able to successfully keep a mouse heart beating in a petri dish for almost a month.

Other organs can exist for periods of time outside the body. Successful organ culture is an exciting and relatively recent addition to the arsenal of biological tools available to scientists. But what do the relationships of organs to the body and their experimental manipulations have to do with understanding the dramatic difference between life and death? We would not argue that the mouse who donated that heart is truly dead. Is the question about whether the organ is alive, beating in the dish, a matter of semantics? Must the definition of biological life be modified to accommodate the ability to culture organized tissues? It is not enough to say that these organs hooked together are miraculously hardy and exasperatingly specific. The individual organ, separated and distinct, also exhibits life. To understand how that relates to a definition of biological life, we must delve into the interior of these organs. That really means examining their individual building blocks, or their cells.

The Art of Celling Yourself

Let's begin by talking about a typical cell, an entity so specialized that the designation is an oxymoron. For our purposes, a typical cell looks a bit like a fried egg, cooked over easy (see fig. 2.1). The yolk of the egg corresponds to the nucleus, and the surrounding egg white corresponds to the cytoplasm. Most creatures that you and I routinely think of as real live organisms are composed of cells. Your body possesses between 20 and 60 trillion of them, collectively working together to confront you every morning with your biological needs and personal capabilities.

This cooperative bunch is governed not by a democracy but by a rigidly controlled biochemical dictatorship. The human being is a relatively weak organism. We have very little hair. We have no protective shell. Our skin is soft and easily punctured, and we can reproduce only once every year or so. Internally the body fights for its survival every minute of every day against forces that often make it sick, and when we die, will consume it. Rigidly defined cooperation and forced labor are the rule between constituent cells in any one particular organ and between organs within a single creature. Specialized cells are under exacting bio-

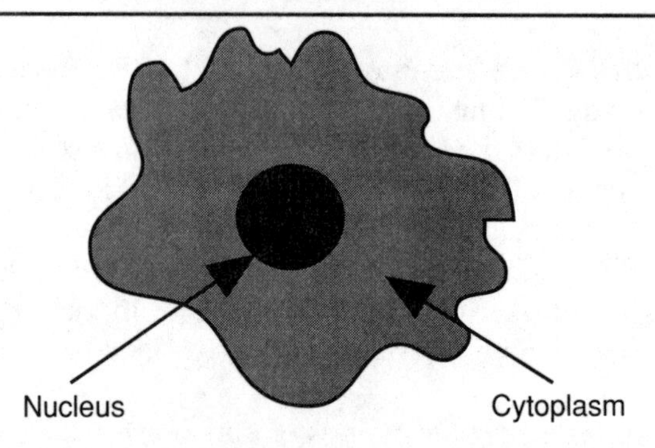

***Fig. 2.1.* A typical animal cell**

chemical authority in the single effort to survive for as long as possible against the forces of a hostile environment.

So how does the body collaborate with itself to fight this war? Some cells have taken cooperativity to such an extreme that they have given up any sense of individual identity. An example is skeletal muscle, a tissue consisting of cells that have fused their many cytoplasms together to form giant, elongated tubes (see fig. 2.2). These tubes contain molecules that can grab onto or slide past each other; this sliding forms the basis for the motion of a typical muscle. If coordinated movement is to be achieved, thousands of these tubes must slide their molecules in a particular direction like tiny oarsmen rowing an ancient Greek galley. Since human motion is a lot like a lever, other fibers on an opposite muscle must move in an equally coordinated but opposing direction for motion to be achieved. This is cooperation at its summit. It happens whether the muscles are in your arms, your legs, or your eyes moving back and forth as you read this page. The importance of this coordination is realized by observing its lack in such devastating diseases as multiple sclerosis and muscular dystrophy.

Cooperation also involves communication between unlike cells. The order to move a particular region of muscle is directed by a neighboring nerve cell, which looks a little bit like a broom with a very thin handle (see fig. 2.3). The bristles at the edge of this cellular broom spit out chemicals that can be received and interpreted by the muscle cell. Thus, local communication between unlike cells at the nerve/muscle junction explains why you can get up in the morning. Those chemicals are spit out in response to a signal originally generated in the brain, which means that communication occurs over long distances, too. A typical design feature that is intercellular coordination is elaborated throughout the body at both short and long distances.

Another great design is specialization. Too many survival problems must be solved for a complex organism to endure on this planet without

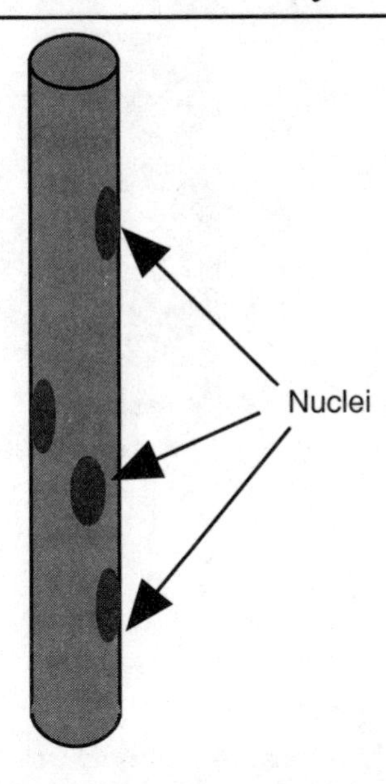

Fig. 2.2. **A typical multinucleated skeletal muscle fibril**

specialization. Take contemporary human digestion, for example. One of my favorite foods in the world is pizza. But eating pizza would be useless from a nutrient-gathering point of view without an incredible array of specialized cells. My digestive system's task is to break apart all those lovely calories into constituent biochemicals so that I can be nourished. Special acid-secreting cells in my stomach perform this service. However, the acids that must be present to do this job are also fully capable of burning most of my other tissues. I have cells that surround these acid producers and do nothing but protect my body from harm. These protector cells will burn away if the acid is unceasingly present, so I have other cells that act as a timing device to stimulate the acid production only at certain times. And then I have to decide what those times are going to be, so I possess cells that can create and sense and respond to hunger cues. The more sophisticated an organism becomes, the more specialized cellular support is required. One of the mysteries of life is how such ordinary-looking cells like a sperm and an egg can fuse to create the incredible cellular diversity of a human being.

The two fundamentals that allow human life to survive the rigors of this planet are thus cellular cooperativity and specialization. The degree of specialization has progressed so far that some cells lose a lot of their cooperativity and become solo performers inside the body. Some cells in your bloodstream can roam almost anywhere in the body. Called *macrophages,* these cells need this access to seek out and destroy foreign organisms wherever infection might occur. Germ line cells, the sperm and

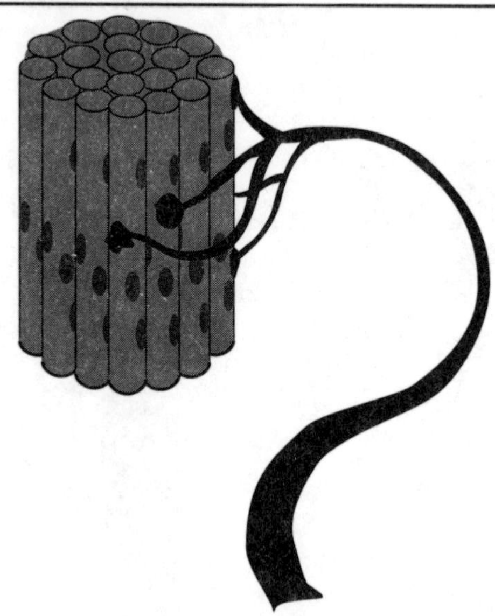

Fig. 2.3. **Connection between a nerve cell and a bundle of muscle fibers**

eggs of the reproductive tract, are so specialized that they have lost any function inside the bodies of their respective hosts. All these types of cells cooperate with the whole: macrophages keep the body healthy, and germ line cells ensure future generations. But their specialization has caused their cooperativity to exist in a different and more independent form than most cell types.

Many of the cell types mentioned can grow independently in dishes within the modern genetics laboratory. We routinely strip cells from individual organs and place them into plates filled with nutrients and buffers. The plates are then stored in an incubator kept at a balmy 98° F. These cells are alive in the sense that they retain the mechanisms allowing them to both reproduce and perform specific biochemical functions. But the organs from which they are derived are no longer in a recognizable form. After stripping an organ, we usually discard what is left of it in the trash can or in a container to be sterilized and thrown away. Thus, it is possible to sustain biological life in the absence of the organisms themselves and even in the absence of the supporting organs from which they originated.

The ability of many cell types to exist for long periods of time in culture has been useful in medical research. Most data concerning human cell biology have come from cell culture since the experiments are impossible or illegal to perform on persons. But what do specialized groups of human cells and their experimental manipulations have to do with understanding the dramatic difference between life and death? We would not argue that the stripped organ that donated the cells is truly dead. Is the question as to whether the cells reproducing in the dish are alive a matter of semantics? Must the definition of biological life be modified to accommodate these abilities? It is not enough to say that cells existing as part of a complex whole exhibit incredible cooperativity and specialization. The individual cells, growing and reproducing independently in petri dishes, also exhibit life. To understand how that relates to a definition of biological existence, we must examine the nature of the various structures and individual chemicals in the interior of the cell. That really means probing the fountainhead of their origins, the genes and the chromosomes of human beings.

Gene Cuisine

If you had the eyes of an atom and could peer inside a typical human cell, the first thing you would notice is a large structure dominating its interior. This is the nucleus, a ball-shaped entity astride the middle of the cell like some ancient Oriental palace suspended in air. Similar to the brain, the nucleus exerts vast controlling powers over the cell. It can export instructions in the form of chemicals that serve notice to the rest of the cell to perform some function. When messages are received from outside the cell, a chemical emissary must often deliver the message directly to the inside of the nucleus. The nucleus has specialized portals of entry, allowing for chemical inspection of incoming and outgoing information (see fig. 2.4).

The molecules that rule from inside this chemical palace are the familiar chromosomes. Chromosomes are made of a chemical termed *DNA,* which under visual scrutiny looks like a rubber ladder twisted at one end (see fig. 2.5). DNA is a thankfully short abbreviation for *deoxyribonucleic acid,* a molecule that is part sugar (called a *ribose sugar*) that has lost a critical oxygen atom (and thus deoxy).

This nuclear palace is extremely crowded. More than a yard of DNA has to fit within its perimeter, a space fractions of an inch in diameter. That's like taking 30 miles of gold thread and stuffing it into a cherry pit. And all this coziness must be in a highly ordered, precisely arranged fashion if it is to carry out the executive functions of biological complexity.

Chromosomes encode genetic information. Lots of it. If all the encoded information were written on a standard typewriter, you would create an encyclopedia fully three hundred volumes in length with five hundred pages per volume. Our bodies condense all of this information to forty-six volumes, the number of chromosomes in a typical human cell. The information inscribed on those chromosomes constitutes the traits that specify the attributes of a single human being. They aren't the actual traits themselves, any more than a book about the Chinese Emperor Han is the

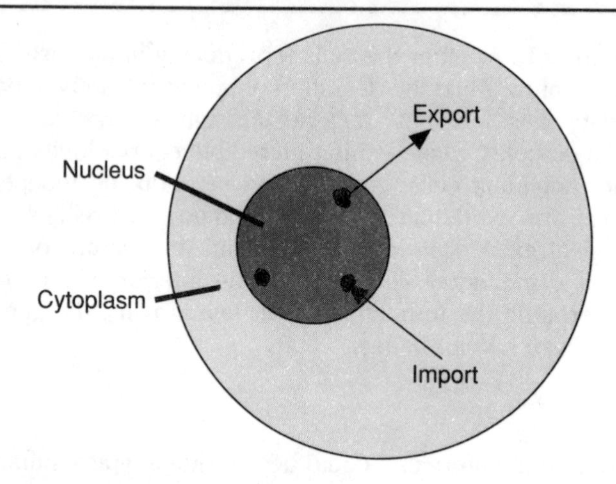

Fig. 2.4. **The nucleus exerts biochemical control over many cellular processes, including what is imported to it and exported from it. The portals for entry are termed *nuclear pores.***

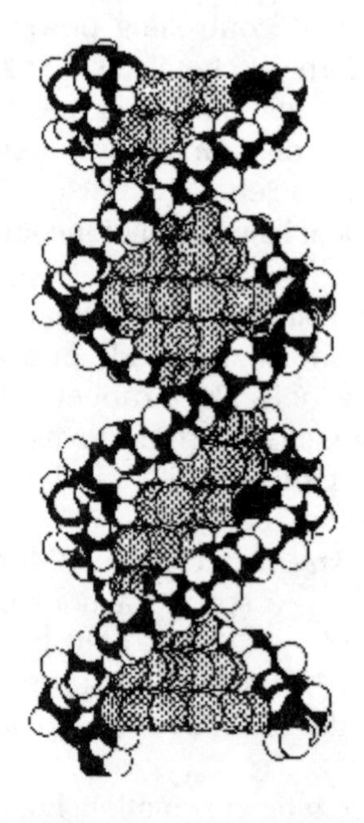

Fig. 2.5. **The DNA double helix**

actual flesh and blood of the monarch. Chromosomes are a biochemical record, a usable template, upon which the schematic for a human life is inscribed.

Chromosomes are subdivided into discrete units known as *genes.* Genes are differentially active parts of a chromosome that exist all along the giant strand (see fig. 2.6). Genes are the fundamental genetic unit, and in times past, were called *traits*. Many such traits are scattered throughout the nucleus; a typical chromosome has three thousand activatable genes. Curiously, a lot of the chromosome appears not to be needed for traits to be encoded and passed on. In a fit of arrogance, we have called it *junk DNA*.

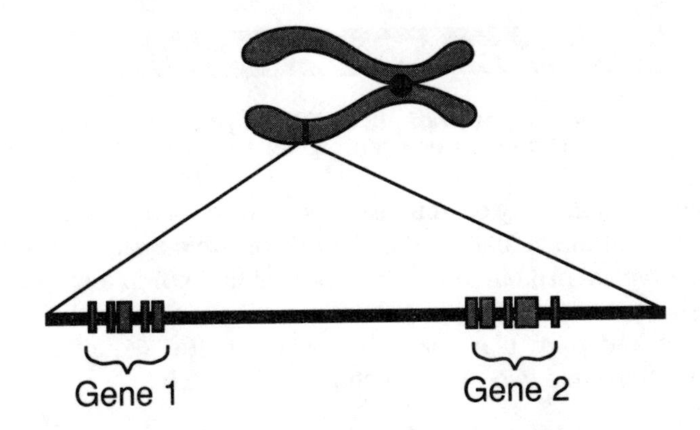

Fig. 2.6. **Chromosomes are divided into discrete units called genes. Note that each gene is itself divided into specific units.**

What does a gene encode? A gene encodes instructions to make a protein, the same type of familiar substance in peanut butter. A gene is composed of small biochemical units called *nucleotides,* like tiny beads on a long twisted necklace. There are only four types of nucleotides in any one gene, represented by four letters: A, G, T, and C (see fig. 2.7). A gene is typically thousands of nucleotides in length. The same four nucleotides are used throughout its structure, regardless of its function or length. Figuring out the order of the nucleotides in a gene is a routine task of the modern molecular biologist. Discovering that it could be figured out at all was one of the greatest triumphs in the history of biology.

The Best Reason for Eating a Peanut Butter and Jelly Sandwich

DNA possesses only the information necessary to construct proteins. That doesn't sound like a big task—hardly the genetic information to make a whole person, who consists of more than peanut butter–related molecules. So what do proteins do? How is it that by encoding proteins, DNA is called the blueprint for the manufacture of a single human being?

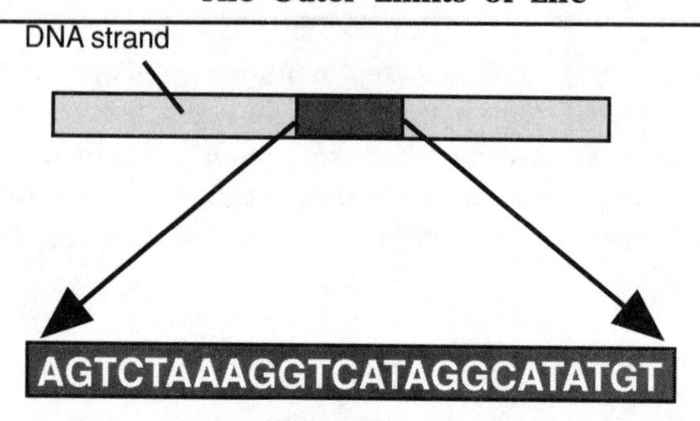

***Fig. 2.7.* A gene is composed of nucleotides, symbolized by the letters A, G, T, and C. The upper box represents a single strand of DNA.**

The answer comes by examining the structure of the proteins. Proteins are made up of individual building blocks, or *amino acids*. Three-dimensionally, a typical protein looks like a pretzel that got in a car wreck (see fig. 2.8). It is a contorted mass of hundreds of amino acids twisted into a very exact and particular shape according to the sequence of amino acids. Proteins have two main functions in the body.

***Fig. 2.8.* A typical protein**

They have a structural function. As you probably know, muscle and hair are made out of protein. Without this biochemical, we would be hairless, immovable blobs without fingernails. Actually, we would be much less than that. Your individual cells contain a miniskeleton just like your body contains a giant skeleton. This miniskeleton is composed of protein, and without it, all cells would cease to function. Proteins are involved in structural integrity, both in the external anatomy of the whole organism and inside the tiny environment of the cell.

Proteins also have a catalytic function. *Enzymes* are a very particular class of proteins. Enzymes perform many vital functions in the body, from controlling how and what we think to controlling how and what we digest. The very ability of individual cells, and by inference whole organisms, to reproduce themselves depends on these proteins. Without enzymes, there would literally be no biological life as we observe it in its present form today. Their function includes catalyzing reaction between chemicals that are not proteins themselves.

Proteins are thus essential in the structure and the function of living creatures. By coding for these proteins, the DNA in a very real way codes for all the information necessary to keep organisms, including human beings, structurally organized and functionally alive.

Molecular Morse Code

So how does a gene make a protein? When we examine the locations of genes and proteins, we immediately see a problem. A gene sits there in the throne room of its nucleus. Proteins, however, reside in both the nucleus and the cytoplasm. Some proteins, like your hair and your fingernails, get squirted out of the cell and into the real world. How does the information in a gene get out of the nucleus and into the rest of the world?

Obviously, the DNA needs a messenger that can precisely copy its instructions, escape from the nucleus, and live to herald its information to the cytoplasm. And that is exactly what the cell does. The messenger is made up of a chemical called *RNA*, or *ribonucleic acid*. In structure it is similar to DNA, also consisting of a sugar (a ribose sugar) that kept its critical oxygen (and is not deoxy). RNA has a property not shared by the DNA in the gene, however: it does not have to stay in the nucleus. It is free to move out into the cytoplasm. There are several classes of RNA, and the one that delivers the message of the gene to the outside is *messenger RNA* (often written *mRNA*).

So how does a gene make an RNA molecule? Two major teams of biochemicals are involved. The first team consists of single RNA nucleotides, the building blocks of the message. These RNA molecules float, like spilled alphabet soup in a space shuttle, all around the nucleus. The second team consists of a sticky aggregate of proteins that wander, also untethered, around these nucleotides and around the various strands of chromosomes. We have termed that collection of proteins *RNA polymerase*.

This aggregate has three functions. Its first function is to grab onto the DNA in the chromosome like a baseball player grabs a bat. Once anchored, its second function is to move along the chromosome, locate a gene from all the junk DNA, and read it. This means moving along the DNA strand and figuring out the order of A's and G's and T's and C's in the gene. The third function is to locate and ensnare those individual RNA nucleotides floating around the nucleus. As the RNA polymerase ratchets along the DNA, it snatches the RNA molecules one at a time and puts them into an exact complementary order of the DNA that lies beneath it. As it places the RNA nucleotides into the order dictated by the gene, it also hooks them together into a single line. Soon a precisely duplicated record of the DNA is constructed, now encrypted as RNA. This RNA can get out of the nucleus and make its way, instructions buried in its chemical interior, to the cytoplasm (see fig. 2.9).

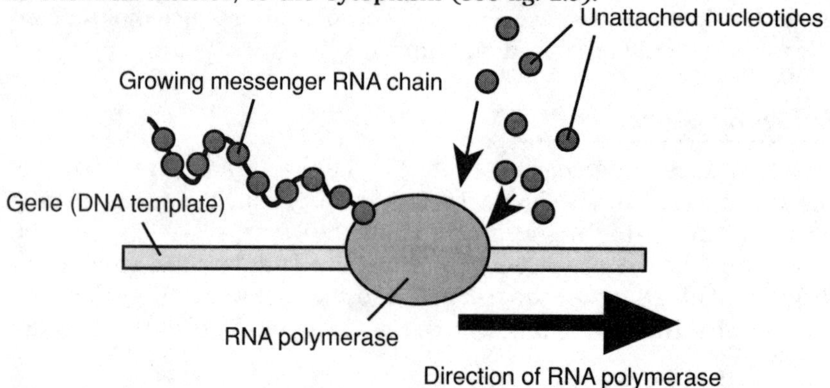

***Fig. 2.9.* Unincorporated nucleotides are processed into messenger RNA via the enzyme RNA polymerase. Using the gene (DNA) as its template, RNA polymerase makes this message so that the gene's instructions can be transported out of the nucleus and eventually made into protein.**

Not Getting Lost in the Translation

The RNA gets out of the nucleus. Now what? The goal is to make a protein. The RNA does not travel very long in the watery cytoplasm before it finds a structure named a *ribosome*. A ribosome looks more like a giant meatball on top of a Frisbee than a sophisticated biological entity. In reality, it is a mobile protein assembly plant and is surrounded by thousands of individual amino acids, the building blocks of proteins. These amino acids are under molecular escort, each shepherded to the ribosome by a molecule known as *transfer RNA* (tRNA, for short). There are over twenty varieties of amino acids, each under the control of an individual tRNA.

When a ribosome encounters a messenger RNA, it finds the beginning of the messenger RNA and reads its encoded instructions. In function, this reading is similar to that of the RNA polymerase back in the nucleus. The ribosome ratchets along the RNA just like the polymerase ratcheted along the DNA. In response to this reading, the amino acids, with prompting from their tRNA consorts, go inside the ribosome. Which amino acids go in? The ones specified by the nucleotide order in the message. The whole point of the genetic code is to see that amino acids are placed into a proper sequence by the ribosomes. The ribosome stitches the amino acids together one by one as the tRNA chaperones escort them inside.

When there are enough amino acids in a row, a functional and specific protein is created. The information in the gene is thus transferred via messenger RNA into a functional protein (see fig. 2.10).

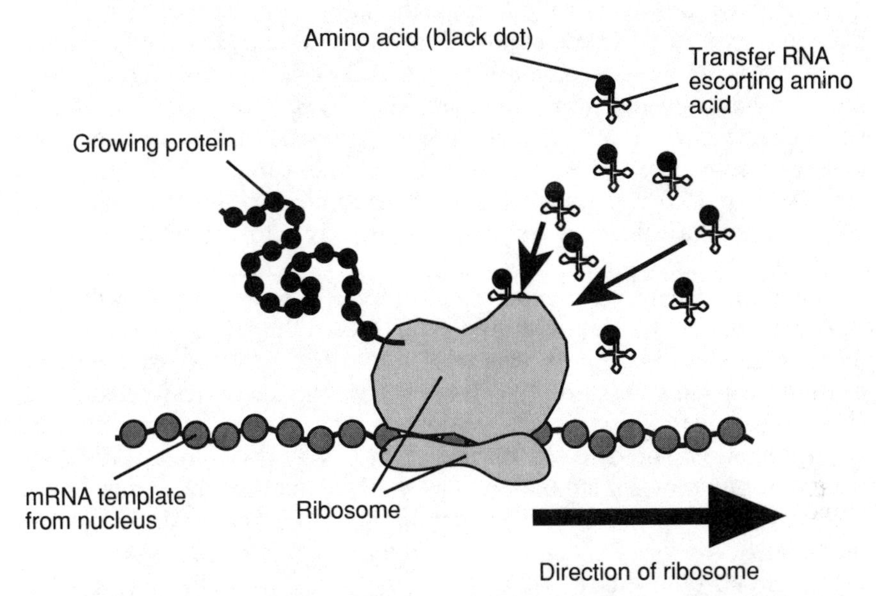

***Fig. 2.10.* Unincorporated amino acids, escorted by individual tRNA molecules, are made into proteins by the ribosome. Specific amino acids are placed into a growing protein according to the instructions encoded in the messenger RNA.**

Flatteries Not Included

The techniques of molecular biology describe how to isolate individual pieces of chromosomes and individual genes from all this complexity. To perform this isolation, we are forced to rip open the cells that carry them, yank the genes out of the nucleus and discard what is left of the ruptured cells. This isolation, though lethal to the cells, in no way disturbs the structure of the gene. The gene is still fully functional.

These techniques have allowed us not only to isolate individual genes but also to manipulate fundamental aspects of their biology. For example, we have learned to reconstitute the RNA-manufacturing process of the nucleus in test tubes. Biotechnology companies sell individual kits that make this process simple and somewhat automated in the laboratory. In a similar manner, we have also learned to reconstitute the protein-manufacturing process of the ribosome in test tubes. We exploit these technologies to synthesize at will the proteins encoded by our previously isolated genes. These artificial proteins behave much of the time like the ones found in the natural world. In such a manner we have imitated some of the controlling genetic functions of a normal human cell without needing the intact cell itself.

What does the ability to create functional genetic units in test tubes have to do with understanding the dramatic difference between life and death? We would not argue that the cells that donated the isolated gene, once we have finished with them, are truly dead. Is the question about whether the gene is alive, creating RNA and protein in a test tube, a matter of semantics? The isolated gene exhibits some of the most intimate and basic properties of life when examined in the laboratory. It does not need a cell, let alone an entire organism, to exhibit its original function.

This point is extremely important. We have seen that it is possible to isolate organs out of particular creatures, discard the creatures, and keep these organs alive. We have also seen that it is possible to isolate cells from these organs, discard what tissue is left, and keep these cells alive. Now we see that it is possible to isolate a gene from a cell, throw the rest of the cell away, and discover that the gene is fully operational. Must the definition of biological life be modified to accommodate these new abilities? Is it possible that the difference between life and death is not so dramatically defined when one considers the murky complexities of genes and chromosomes?

To understand the relevance of these questions, we need to understand exactly how functional isolated genes become in other contexts. We must examine their behavior when placed in cells other than their source or, better, when placed in creatures never designed to accommodate them.

The Psychological Hazards of Eating Alpha-Bits

When I was little, my mother told me that all creation was made of little particles that existed, a lot like me, in a constant state of motion. She also told me that my entire biological existence was controlled by tiny clusters of these particles. I told her that this could not possibly be true. When I looked in the mirror, I did not see any of the tiny bumps on my face that surely must be present if I was made of little particles. Moreover, I could lie perfectly still on the couch and not feel my insides jerk around like jumping beans.

My skepticism did not waver as I grew older. In science class, I was told that my complete genetic makeup was composed of those four molecular letters arranged in a certain order. I immediately interpreted the teacher's statement to mean that we were all made out of Alpha-Bits cereal. I became terrified when I learned that if those same four letters were placed in a different order, they would describe an eggplant. Or a fish. Or the virus that gave me colds every winter. So I stopped eating Alpha-Bits cereal, thinking that if one of those letters got loose in my body, I would turn into a snail. Or a girl.

It was a long time before I became convinced that biology was carved from genetic building blocks, just the way my mom, my teachers, and textbooks said it was. My disbelief then gave way to a sense of wonder followed, as I grew older, by a dull sense of warning. It became clear to me that genes were the absolute and final control point for defining biological variety. I started to feel that if anybody learned to exploit their power, the individual would possess the ability to manipulate life and, ultimately, to create it.

Those feelings, tempered now by many years of education, have not substantially changed. It turns out that manipulating genes really does yield powerful control over the fate of an organism. As a scientist, I am startled by the ease with which it can be done. Most organisms will accept foreign genes into their genetic structure and replicate the invaders as if it were the most natural thing to do. It is a little bit like discovering a nation whose inhabitants all drive the same kind of car. Entire engines, spare parts, and the tiniest screws are freely interchangeable, and you can mix and match them at will.

Understanding how to isolate and rearrange genes in organisms has allowed us to greatly modify existing creatures. We can transform the functions of whole groups of cells. These techniques have given us the power to snoop into such forbidden topics as aging and attempt the deliberate genetic alteration of human life. Scariest of all, this technology has forced us to realize that because we deal with the molecules controlling every facet of who we are, we deal with life in its most elemental form.

The Splice of Life

An experiment that demonstrates the mixing and matching of genes and organisms involves the firefly and the tobacco plant. All of us have seen fireflies in a summer's evening, floating through the air like little suns in a tiny solar system. Fireflies glow in the dark because of a protein called, no kidding, *luciferase*. This protein facilitates a highly specialized chemical reaction whose end result is the production of light. The gene that makes this protein was first isolated in a most disgusting fashion. A large group of fireflies were collected, killed with chloroform, and then stuck into a blender, creating a firefly milkshake. In a complex series of manipulations, the luciferase gene was isolated. The gene was then placed into the cells of a tobacco plant through a standard electrical process. When the tobacco plant grew, it replicated this foreign gene along with the rest of its chromosomal material. And when the tobacco plant was fed certain nutrients, the plant glowed in the dark, just like a firefly.

Despite the huge biological difference between the free-flying beetle and the sedentary plant, their genetic machineries exhibit enough cooperativity that the plant can support the presence of a specialized gene and drive it as if it were the insect itself. And the luciferase gene contains all the information necessary to collaborate with much of the plant's machinery and behave as if it were in a beetle. The end result is that the life of one insect can be transferred to the life of a plant, a fusion we used to think privy only to eggs and sperm. This example demonstrates that even the most specialized genes can cooperate with the most unlikely organisms if handled in a certain fashion.

This specialization and cooperativity has been exploited in a number of ways and for a variety of reasons. We have placed goat genes into mice and created little rodents that secrete goat's milk. We have created giant fish by supplying little ones with an unrestricted mammalian growth hormone gene. We have successfully implanted part of the human immune system into laboratory mice. Curiously, they are beginning to grow human organs. We have constructed two-headed frog embryos. We have installed pesticides directly into the genetic machinery of plants so that we do not have to spray them. We can do these things because most genes have an incredible ability to cooperate with the internal biochemistries of most foreign organisms. We have the frightening potential to manipulate any creature on the face of the earth.

For Whom the Cell Tolls, Part I

The fluidity of genetic transfer has been useful in more than redesigning organisms, however. It has also been used to explore the delicate genetic mechanisms of single animals and the even more delicate mechanisms of individual cells. This research has yielded insights into why certain biological tissues are implacably sturdy and why others have finicky requirements for growth and development.

For example, an organism used a lot in genetic research is the lowly fruit fly, *Drosophila melanogaster*. The development of what is often termed the body plan has been studied in great detail, including how and where this insect gets its legs and wings and antennae. It was found that limb development is normally controlled by certain genes. There is a gene called *ant*. If you knock out the function of this gene, a leg instead of an antenna will grow out of the head of the poor fruit fly. Arresting the development of an entire appendage like an antenna by knocking out one gene was curious. But also promoting the growth of a leg in the antenna's usual spot was a complete surprise. It was a little bit like throwing a hand grenade at a sailboat and having the resultant explosion turn the boat into a battleship.

You can knock out the function of another gene called *sxc*. This will cause the unusual leg/antenna swap and will also cause the wings to transform into an organ the insect usually uses as a gyroscope in flight. This poor mutant creature has a leg coming out of the middle of its head and has incredible navigation skills but it cannot fly.

These experiments demonstrate the mysterious interconnections that genetically wire organisms together. These limb-controlling genes—termed *homeotic genes*—reside in humans as well. Although humans cannot grow antennae, do not possess in-flight gyroscopes, and cannot fly, we do possess a body complete with legs and arms. It is hoped that such research will furnish insight into how humans develop a body plan and how single genes can orchestrate the activities of whole batteries of other genetic instructions.

For Whom the Cell Tolls, Part II

We have been able to extend this manipulation beyond destroying the functions of specific genes. With the techniques of molecular biology we have been able to change the very nature of the cells themselves. Most cells, like most organs, have narrowly defined requirements for function. Your adult fat cells will never become adult nerve cells, your eyeballs will not grow teeth, and your ankles won't sprout earlobes. Each cell has its own requirements for growth and its own specialized functions in the mature adult. We have known for a long time that such specialization did not always exist in the life of a human, however. As an embryo, every human possessed at one time the genetic ability to convert simple watery precursor cells into not-so-simple tissues like bones and muscles and specific organs. One hypothesis was that a class of molecules so-called the master genes existed. Such genes would be powerful, working like a conductor of an orchestra to signal whole sections of the embryo to differentiate and specialize.

Master genes have been isolated from human tissues. One such gene, first isolated by Bob Davis and Hal Weintraub, is called *MyoD1*. When MyoD1 is genetically stitched into the proverbially unwanted fat cell, the fat cell begins to elongate, make internal muscle fibers, and fuse with neighboring cells. Under the direction of the MyoD1 gene, the fat cell turns into skeletal muscle. This conversion function is powerful. MyoD1 has been placed into large numbers of different types of cells and can convert almost all of them. MyoD1 represents a whole class of muscle conversion molecules, but it is not the most powerful member. These genes can break apart the finicky differentiation of mature cells and reprogram their genetic fate.

The ability to control the fate of cells through switching master genes on and off gives us the potential to manipulate the genetic constituency of embryonic organisms. However, the insights gained by such research do not apply only to the beginnings of life. That master genes also control the endings of life is becoming increasingly obvious.

Pushing Back the Glands of Time

Let's examine the aging process as an example. Aging is a deterioration we draw molecular swords with the day we are born. Molecular biology is learning about this unwelcome decline by investigating the following paradox. We have known for many decades that individual cells replicate by dividing into exact replicas of themselves. One cell splits into two identical cells through a process known as *mitosis*. But the cells are finicky. They will split for only so long before they will divide no longer. Instead, the nondividing cells will undergo the process of *senescence,* and then they will die.

To keep functioning, your body has to manufacture new recruits to take the places of their fallen brethren. You get a nearly complete turnover of soft tissue cells every eleven months or so (your skeleton is a little slower, reproducing every seven years). Every year, you get a completely different body from the one you had the year before. The paradox is this: if most cells divide identically through mitosis and you get a new body every eleven months or so, why do you age? A new identical body should reappear year after year with very little change. Aging, as we know it, should not occur.

As our skin, muscles, and bones testify to us, especially as we get up in the morning, we do age. Clues about why we age come from research done in tissue culture. You can place the cells of human beings into a petri dish, and many of them, if given the right nutrients, will live for a while. But they almost always die. All except one type. Many types of cancer cells, if supplied with the proper nutrients, will never die. It is an example of hardiness taken to such an extreme that scientists use the adjective *immortalized* to formally describe the phenomenon.

We work with cells in the laboratory from a woman named Helen who died of cervical cancer in 1956. Her tumor cells are growing like gangbusters in culture. They are so hardy that if we don't take precautions, they will rise up into the air and contaminate other cultures. Researchers have concluded that the clue to the aging process must be somewhere between normal cells, which like us always die, and cancer cells, which unlike us can be immortalized.

The answer is beginning to be unraveled, and the implications of this research are the stuff of science fiction. Scientists have discovered that the aging mechanism many cells exhibit is the result of a particular gene or series of genes. This mechanism is located on a part of human chromosome number one. This evidence demonstrates that planned obsolescence is a trait genetically prewired into most cells. At prearranged times, the cells are told to no longer divide; they are told to quit, to senesce, to age.

Why is this interesting news? As seen with the fruit fly, it is possible in the laboratory to knock out the function of any particular gene of interest. When those control mechanisms on chromosome one are understood, curbing the aging process in cells by destroying the genes responsible for the degeneration may be possible. The exotic hope is that if we can control such processes in cells, we might eventually control them in organisms.

The conclusion of all this work? These genes are little bits of life that contain the most powerful information available to biological organisms. Such profoundly intimate influence has allowed us to mix and match traits in creatures just as one selects a wardrobe at a department store. We use the genes to explore the delicate biology of entire organisms and convert finicky cells into specific types. Not surprisingly, we have found these molecules to be intimately involved with the endings of life as well as the beginnings. That brings us back to our original question about the dramatic differences between the states of life and death. If we can control the biology of the genes, we can control the life of the organism. In fact, the biology of these genes may be the definition of life in the organism that carries them.

A conclusion that profoundly alters the question.

Heartbreak

A tough task faced by an emergency room physician is telling anxious relatives that a loved one has died. The job is made even tougher if the person who died did not leave any information about the wish to donate organs. Then the physician must seek permission from the bereaved relatives to harvest the organs of the dead person. It's so hard because the last thing in the world shocked and grieving people generally want to deal with is organ donation. Nonetheless, if the organs are not removed as quickly as possible, they will not be in good enough shape to be useful in transplants. The survival of potential recipients of the organs depends on the ability of these relatives to peer past the death experience of a loved one and give permission for organ harvesting. It is an odd thing that a dead person has enough life to save another who does not have enough life to save himself or herself—as long as the living give their permission.

I remember a colleague's experience in his residency. He was in the intensive care unit of a research hospital when he heard the ward's phone ring. The single occupant of the unit was a forty-five-year-old plumber, fighting for his life amidst the machines clanging a noisy narration of his condition to anyone who passed by. The man suffered from idiopathic dilated cardiomyopathy. In this disorder, the heart slowly rots away like old cheese in a refrigerator, and blood backs up into the malfunctioning chambers and can stretch the heart to huge, unworkable proportions. He desperately needed a new one. And the family members with him desperately needed to hear that a donor had been found. By the time my friend picked up the phone in the unit, the man had probably no more than ninety-six hours of life left in his body.

The phone call was from a physician at a distant hospital. Around 4:30 in the afternoon, a twenty-one-year-old male had been wheeled into the emergency room of the hospital. He had given himself a gunshot wound to the head. His heart was functioning, but his brain had very little electrical activity; he was what is often termed a *heartbeating cadaver.* The parents of the man were right behind the ambulance and were placed into an adjoining waiting room. Because of the young man's condition, a representative from the organ donors' association was called, as was the hospital chaplain. Along with the attending physician, they informed the anxious parents of their son's death and then asked for permission to harvest his vital organs.

In a way the request seemed odd. The dead man was not mangled or misshapen. The way his head rested on the pillow, he could have been mistaken for being asleep. Nonetheless, he was dead, and the parents in the midst of their grief agreed that their son's organs could be harvested. The mother said through her tears what many bereaved people say in response to the request: her son would have wanted it. She felt that a part of her son would live on past his death, doing what he loved to do best, which was making people happy. And then a phone call was made to inform a team of physicians at my colleague's hospital that a donor for the forty-five-year-old plumber had been found.

Life in Death Situations

At this point technology has a head-on collision with the human experience. Transplants can occur only because a dramatic life on/life off switch does not exist in complex organisms like humans. People are made of individual components, each containing biological functions capable of limited independent survival in the clinic or laboratory. Thus, just because the entire organism is no longer functioning as it used to does not mean that all biological life has disappeared.

This fact is not particularly popular. Many of us would like to think that biological life is generated by some God-given mysterious and unmeasurable ether that completely dissipates when an organism dies. Physical life may be God-given, but it is not necessarily mysterious. We cannot always take the life of a particular organism and find all of its functional biochemistry subsequently destroyed. If that were true, physicians would never be able to transplant organs, and scientists could not support them in culture.

Happily for the recipient, we can transplant organs. A national bureau is set up to find and match recipients with available and eligible donors. A great deal of activity always accompanies news of a successful match. In the case of the plumber, a team of medical personnel were hastily assembled to go to the distant hospital and harvest the waiting organ. From extraction to implantation, everything had to be done quickly since the heart is only in usable condition for about five hours after it is removed. The general rule for most physicians is that the recipient can be sedated but not cut unless the donor heart is in the same operating room. So while the extraction team ran out the door of the hospital, the plumber was prepped for surgery.

At the suicide victim's hospital, the members of the extraction team were hurriedly ushered into another operating room. Like surrealistic vultures of mercy, other groups of doctors were already at work on the body of the young man, busily extracting various organs. Since the heart must supply nourishment to the rest of the tissues, it is usually the last to be taken. When it was his turn, the heart surgeon injected the aorta with a paralyzing solution to stop the heart from beating and then removed it from the donor's chest. With great care, the surgeon placed the living heart on an examining table and peered inside to evaluate its fitness for transplantation. After he was satisfied that there were no concealed problems, the heart was quickly placed in a jar filled with precooled chemicals and whisked out the door. The second team drove as fast as they could back to the operating room where the anesthetized plumber lay waiting.

Law-Inspiring Biochemistry

It has been not been very long since the teams under Christiaan Barnard performed the first heart transplant. Since that time, the procedure has been modified and improved. Such manipulations, like all medical procedures, are doable because biological processes are based not on sorcery but on determinable and predictable physical laws. As has been stated, when we crack open the biology of a cell and look at life, we do not find murky metaphysical forces holding the generators of existence in an unfathomable grip. Instead, we observe a miraculously ordered biochemical universe made all the more magnificent because it is so small. We find the hardiest of organs pampering fluids that are as delicate as they are finicky. We find almost slavishly cooperative team players locked in a molecular concerto with free-ranging cellular soloists. We see bits of life in cells that flicker their own signals like a lighthouse to neighboring cells and to far-flung tissues.

This world is so ordered that we can design experiments and make predictions based on their outcomes. The realm of biology is a celebration of God's creativity in its most beautifully miniaturized form. There is no violation or shame in taking away this mystery. Like pulling curtains from a window in a mountain cabin, the joy is that the view we behold is so spectacular.

The team with the donor heart arrived at the hospital and again was rushed into an operating room. Because it received no blood and thus no oxygen in its glass jar, the heart had already turned an unhealthy blue color. It was also rapidly losing its ability to remain transplantable. The surgeons hovering over the plumber very quickly cut into his chest and hooked him up to a heart-lung machine. The machine would gradually replace the functions of his own heart and lungs while the new organ was installed. After a series of clamps were positioned, the plumber's old heart was removed. Eventually, the new heart was placed into the chest, the proper vessels hooked into place, and the patient administered his first dose of a lifelong series of anti-rejection drugs.

God's Exchange and Return Policy

The patient's old heart looked more like a sagging gray vacuum bag than a member of his anatomy. It was treated as most of these organs are treated, discarded as a dysfunctioning relic of a future no longer threatened. However, for molecular biologists, the organ is infested with life. If we were interested in understanding why the cells in the man's heart failed him, we could, with his permission, obtain the organ for study. We would place some of the heart's tissue in a blender, extract the DNA sequences, and put them into tiny viruslike particles. Copies of his genes, still fully excitable, would then be available to isolate and clone at will. In a fashion, the heart would live on separated as individual pieces in a hundred test tubes.

These kinds of biological studies and the techniques of organ transplantation illustrate the human side of a disturbing fact: life is sustainable by the exchange of internal biological parts. In the real world, these technologies affect living, loving people like you and me. We at this moment possess tissues that are of great potential benefit to others if circumstances cause us to donate them. We also possess organs capable of becoming as dysfunctional as the plumber's heart. This fluidity is uncomfortable because we are not inclined to think of ourselves as some genetic car repair shop, complete with maintenance requirements and replaceable parts. And we're even less inclined to think about life as a divisible commodity, that the total cessation of the organism does not necessarily equal the total cessation of life. These are the facts, regardless of how we feel about them, however. And they must be taken into consideration if we are going to arrive at a satisfactory definition of life.

What is the dramatic difference between life and death? From a molecular biologist's point of view, there is none. Many of my colleagues would say that the question is irrelevant. We cannot distinguish a dead gene from a live gene any more than we can distinguish a dead carburetor from a live one. Either it functions in certain ways, or it does not. This reasoning also applies to more complex organizations of genes that are packaged as individual organisms. If there are sufficient genetic organization and a friendly environment, a biological entity can perform certain functions. If these requirements are not met, the organism cannot fulfill those functions. The sum total of those processes, governed by the creature's genes, best describes its living state.

If this is true, pursuit of the dramatic difference between biological life and death is a lot like my friend Larry's pursuit of the totally American car. Investigation of the answer has changed the nature of the question. Organisms do not possess a mysterious, metaphysical on/off switch forcing every resident biochemical capability to come to a screeching halt once the organism has died. We instead find gradual and predictable biochemical reactions at work in all living things. They even obey the laws of chemistry and physics. This brings us to a disturbing conclusion. Biological life may be only one end point in a vast continuum of biochemical processes. To investigate the nature of this kind of life we will be forced to ask a different set of questions. And we will be forced to rethink some of our definitions, not only about biological life but about biological death.

It is the subject of the next chapter to find how much rethinking we will have to do.

Chapter 3

Circular Seasoning

Introduction

As a little boy, I tried to get a garter snake to swallow its tail. I went out to our backyard garden and found a long yellow-and-black garter snake. The first thing I tried to do was to push that long tail into the poor snake's mouth. That didn't work. So I put the snake into a glass jar and went back to the house to retrieve, of all things, a jar of peanut butter. I reasoned that if I smeared what was to me a universal gustatorial delight on the back end of the garter snake, I could get it to swallow its tail. So I got the peanut butter and smeared a little bit on its tail. No luck. I smeared a little more on its tail. Still no luck.

Finally, I smeared the peanut butter all over its body. The snake was so slippery that I could no longer hold onto it. It escaped from my hand, probably thinking how wrong biologists were to conclude that humans exhibited the highest form of terrestrial intelligence. I was left with a shirt full of peanut butter, wondering why I could not get the snake to exhibit a behavior that I deemed natural for it to perform.

I would like to use this peanut butter mess to redefine a biological process that the snake has already experienced and that someday we will too. That biological process is death. I use the word *redefine* because the world of molecular biology increasingly shows us that the familiar definitions may be in error. Many dictionaries define *death* as the "absence of life." They also define *life* as the "absence of death." The luxury of this inadvertent circularity is that the issue can be avoided. They are therefore useless as biological definitions. Because they contrast life with death, the definitions imply that one state cannot go on while the other exists. The assumption is that life and death are mutually exclusive biological processes, a subtle, but I believe extremely important, error.

There is, in my opinion, no compelling reason to believe that life and death are mutually exclusive states. The evidence suggests that death is a cumulative rather than an absolute biological experience. This accumulation may show that life and death represent ends of a continuum not of biological absolutes, but of active or inactive genes. It's an old lesson of science. Circular issues are fine if you wish to understand how snakes swallow their tails. The only error in the investigation is first assuming that they can.

Why the Mantis Is Praying

In this chapter I am going to talk a lot about different kinds of animals, beginning with the praying mantis, specifically, the male praying mantis. This gentleman deserves a lot of pity, especially when it is time for him to participate in the reproduction of the species. The male and the female start out exchanging genetic information with the characteristic passion of most animals. However, while locked in her lover's embrace, the female praying mantis turns her head around and begins literally to eat the male for dinner. She starts at his head and continues dining through the neck and torso. Dutifully, what is left of the poor male continues his copulating motion until she finally devours his last, still-moving abdominal segment. That particular segment of the male houses an independent nerve center that is the sole controller of his sexual motion. The male will not commence those amorous gyrations until the female chews his head off. Why? His head contains a chemical inhibitor of sexual motion. That inhibitor will physically prevent him from completing the act of reproduction until the head is removed. Thus, we have no better fusion of life, sex, and death in the natural world than that of the hapless male praying mantis.

Sexual reproduction is an interesting and surprisingly rare way to carry on the life of the species. In terms of sheer numbers, most organisms don't procreate in this fashion. This is because the vast majority of living things are not visible to the naked eye. They exist in that all-too-nightmarish world of bacteria and other single-celled animals. The bacteria in a single piece of rotting cheese easily outnumber the population of all vertebrate animals on the face of the earth. These organisms do not reproduce by exchanging sperm and eggs.

So how do most organisms procreate? Most reproduce in an efficient, if boring, fashion through the process called fission or mitosis or, in some cases, budding. A cell, say a bacterium, doubles its genetic information by copying its resident strands of DNA. After a short period of time, the bacterium will split into two somewhat equal parts. Each part gets a copy of that replicated DNA stuffed into its cytoplasm. Where there was only one complete organism, now there are two. The parent in this scheme does not have a separate child. Rather, the parent is one of the offspring.

A somewhat disturbing characteristic about this method of reproduction is that these organisms do not have planned obsolescence built into their genetic code. If given enough nutrients, the organisms will continue splitting and dividing forever. They need never experience death, at least not in the way we normally think about it. When these bugs get old, they make a double of themselves and divide. If the environmental conditions are correct, they will live and divide without limit. It may seem extraordinary, but for them, death is not a mandatory result of life. And most organisms have this capacity.

Mating for Life

So who experiences compulsory death? Like our poor praying mantis, only those organisms that have sex must experience compulsory death. Since only more complex organisms reproduce sexually, the inevitability of death is reserved for relatively few creatures.

Sexual reproduction is much different from fission. Instead of one boring and sexless creature, sexual reproduction involves two genders. These genders are made genetically distinct from each other, yet they possess a sexual affinity that allows the distinction to be blurred under certain circumstances. Carried within the bodies of each gender are specialized cells, germ cells, that are unlike any other entity in the biological universe. Alone, these cells have absolutely no function in their respective hosts. They contain only half the genetic information necessary to survive. But this halving of the genetic potential gives rise to the doubling of the species.

The act of mating gets these specialized cells together. As glorious as we have made it, the sole function of mating is to place sperm and eggs in the same general area. These cells have on their surfaces grappling hooks made especially for each. The results of this mating mean that the grappling hooks will find each other and the cells will fuse. This union of cells then triggers a process that will allow growth of a new organism, occurring in the somewhat out-of-focus image of the two parents. When birth transpires, this third organism will be biologically distinct and will carry the information necessary to reproduce itself at a later time.

Why reproduce in this fashion? Why not split like most of the microscopic world? The genetic donation model and the division of species into genders ensure variation of the species. If humans reproduced like bacteria, we would all look somewhat like the first person to have done the split. In other words, we would all be clones of one another. If all people were clones of me, they would have bad eyesight and be under six feet tall. And that is an awful reality in the dark or on the basketball court. Sexual reproduction ensures that humans will survive power outages and that the NBA will have competent centers for centuries. There will be an incredible variation of genetic displays even within a single species if each offspring is a genetic smear of the two parents.

When parents have fulfilled and exhausted their reproductive capabilities over the years, they have a sad but inevitable rendezvous with death. When women stop menstruating, the body goes into a state of natural decline. The lack of estrogen causes bone demineralization. Joints stiffen. Breast tissue begins to atrophy, and there is a general weakening in all the supportive structures of the pelvic floor. The rates of certain cancers go up. Men, whose reproductive capacities can exceed women's by several decades, do not undergo as dramatic a decline. But their ability to reproduce diminishes with age. Eventually, both genders will die, an experience shared with every animal that likewise has a gender. Such a fact probably sends Freudian psychologists and their poets skipping into gardens of metaphor. In the end, the grave is the price we pay for being able to have sex.

The Bugs in the Eiffel Tower

Certain kinds of death, like the Eiffel Tower, are easy to spot because they are so obvious. I remember the first time I encountered a death experience. For most of my growing-up years, my father was in the military. We lived on an air base in Europe that was constantly filled with the almost offensive roaring and buzzing of military aircraft. As I was walking home from school, I noticed a DC-3 lumbering through the atmosphere. I use the word *lumbering* because it seemed to be having trouble staying in the air. When I turned my gaze from it, I heard its engines cut out. I looked back up and saw it cartwheel to the ground like a giant kite. The airplane exploded in a field not one thousand yards in front of me. I was told that eight people died that day, people that my friends and I knew. My first introduction to death was watching that burning wreck smolder and hiss in a field where I could have been playing baseball. It was final, it was awful, and most of all, it was obvious.

As a scientist in a research hospital, I have seen death in much more ambiguous forms. I have seen patients in comas and children born without skulls. I have seen cancer patients whose only noticeable distinction from a corpse was that they could still blink. I have tried to compare these experiences to the plane wreck I first saw on the air base. The only thing I could think of was that death wore a coat composed of many colors. It reminded me of something an old professor of mine told me about scientific investigations. He said that there were two ways to find the location of the Eiffel Tower. You could always first try to find out where it was. If that failed, you would be forced to find out where it was not.

The concept of biological death is easy to observe in cells that are single and free living. When it is no longer dividing, the cell crinkles up like a potato chip in culture, its life span over. But the concept begins to lose its meaning when a creature is beheld as the sum of its parts. In complex organisms, cells are dying all the time. By the time you go to bed this evening, you will have lost untold thousands of your cells. But you are not dead. Even though you experienced very real internal death today, you can read this paragraph. As you recall from the last chapter, your old body will die every eleven months. However, unless you are feeling unusually self-conscious about your birthdays, no one will hold an annual funeral for you every time your cells completely turn over. Thus, biological death holds a curiously cumulative feature about it. The process retains absolute definitions only in narrowly defined circumstances.

To look at the other side of the coin, a dead organism is often full of life. Most animals, including humans, have millions of flesh-hungry microorganisms in their bodies throughout life. When the animal dies, the immune system, which kept the gruesome specks of life in check, is no longer functional. The microorganisms are free to eat at will, multiplying and dividing endlessly. In terms of sheer numbers, the dead carcass is more full of life than it was as a functional animal. Dead creatures can sustain an enormous amount of life even after they have expired. Life-and-death events for complex organisms are not necessarily mutually exclusive processes but can simultaneously occur in various degrees.

Running in the Wrong Circles

It is difficult to obtain an adequate definition of death because it is difficult to obtain an adequate definition of life. Most dictionaries use as their primary definition of death the state that exists when life is not there. But we can continually whittle down complex organisms until we have only their genes left. Some form of functional life is reserved at each reducing step, even if all that remains is a test tube of genes filled with hostile chemicals. My old professor never told me that finding the Eiffel Tower could be extremely perplexing if the monument keeps changing locations.

The point is that defining death by contrasting it with life is not easy. If that is the case, what do we have left? If we go back to the dictionary, this time excluding any references to life, we can distill three common categories of meaning. They are as follows:

1. A lack of feeling or sensitivity
2. An inability to reproduce
3. Lacking animation, inanimate, without normal function

But do these residual definitions give us any better insight into the nature of death? We tend to believe that any organism that can encounter life can also encounter death, even if death is an option. Any definition of death will have to fit within the framework of any creature we say experiences it. We will examine the biology of three simple but very different organisms to test the adequacy of these remaining definitions.

The Graceful Dead

Let us first consider the life cycle of the butterfly. As you know, beautiful butterflies start out as not-so-beautiful caterpillars. Most caterpillars are held to their ugly shape by a combination of internal water pressure and gas, just like a water balloon partially filled with air. If you puncture the outer tissue of a caterpillar with a needle, the body will quickly deflate, and the caterpillar will die. This biology is very distinct from its free-flying destiny, meaning that somewhere down the line the caterpillar will have to be destroyed.

The task of the caterpillar is to eat. And eat. The caterpillar eats so much food that it gains weight at nearly an exponential rate. The internal anatomy of the caterpillar is singly designed to cope with this voracious appetite. Its internal body consists of a couple of nerves hooked up to a very tiny brain. The rest of the caterpillar is one giant stomach. A set of very active genes in the caterpillar keeps its biology in this ravenous wormlike state. They are competing with other genes that would like to become activated and turn this beast into a butterfly. The caterpillar keeps these butterfly genes silent by secreting a substance that performs functions similar to a Peter Pan-turned-Stalin. This substance is called *juvenile hormone*. It physically prevents the activation of genes that would make the caterpillar grow up. Only when this hormone falls below a certain level is the dictatorship broken and the caterpillar allowed to proceed to the next stage of development.

Once the juvenile hormone levels are depleted, the genes that will create a butterfly are free to express themselves. The caterpillar, responding to this new government, weaves its famous cocoon around its body as a combination fitting room and prison. After the cocoon is made, the caterpillar is turned into gravy. Stage-specific biochemicals martial whole battalions of cells to crawl off the old caterpillar form and stand by for new instructions. New cells are refitted to accommodate functions that will allow this creature to fly. The old and new tissues work together to congeal this gravy into the form and function of a butterfly. In the end, a crawling sedentary worm is turned into one of God's most glorious creatures. The whole process can take between two weeks and two years, depending on environmental conditions.

The examination of the life cycle of this insect brings up a problem in defining death as a black-and-white state. No one would argue that a moving and eating caterpillar exhibits life. So does the butterfly. But when this creature is just complex liquid sloshing around in the chrysalis, there is neither worm nor wing. There are no feelings, no instant ability to reproduce, and no real whole-organism animation. Is the caterpillar deceased? The insect has not died in the sense that its potential to flit from flower to flower has been quashed. Nonetheless, while it is unformed, the status of its life under the criteria mentioned is open to question.

The life cycle of this insect points out a fundamental flaw in defining death in black-and-white terms. Such rigid definitions of death are not easily applied to organisms that dramatically change forms. In fairness to the concept, it should be pointed out that there are solid aspects of this insect's life decisions. That same butterfly can never crawl back into its chrysalis and become a worm nor will the chrysalis create a new caterpillar. The life-gravy-life cycle of the developing butterfly is unchangeably one way. And once the butterfly has expired, there is a definite end to its life cycle, and it can only then become fertilizer or food.

But what if you could find an organism whose end wasn't so permanent? What if there was an organism that could live in one stage, die and then reform that stage? What if the organism could perform that transition for centuries, given proper living conditions? If such an organism existed, would the distinctions between life and death blur? Especially if it could form a complex, multicellular beast from single cells? Such a creature does in fact exist, and its life cycle is the topic of our next discussion.

Slime Doesn't Pay

A fascinating organism called *Dictyoselium discoideum* goes by the unceremonious name of *slime mold.* These organisms start their lives as free-wheeling, single-celled, amoebalike creatures. At this stage of their lives they are called *myxamoebae.* These creatures creep along the forest floors of North America, pouring their cytoplasms against the borders of their outer cell membranes like sand into balloons.

The myxamoebae reproduce by doubling their DNA output and then splitting in half. Like so many other organisms, the slime molds will continue this mode of reproduction forever if there is proper food. This fact is frightening to many bacteria, the slime mold's favorite food. Myx-

amoebae can gulp as many as five of these critters a minu
eat another slime mold it encounters on the forest floor.

Cannibalism is not always observed in these creatures, how
the normal food supply is exhausted, some of these myxamoebae will secrete the hormone *cyclic AMP.* This hormone acts like a molecular all-points bulletin, beckoning every slime mold in a given area to the myxamoebae that secreted the signal. In a process known as *chemotaxis,* other slime molds stop crawling in a disorganized pattern and move toward the cyclic AMP. When they encounter each other, they extend the most flattering form of cooperation. They join together and form a single crawling slimy organism. This new organism is termed a *pseudoplasmodia* or, more appropriately, a *slug.* Even though it is composed of many individual cells, this fungal Noah's ark moves as a coordinated unit, having both a front and a back. It leaves a slimy trail wherever it goes.

The slug is under intense scientific investigation because it represents the crudest form of a whole organism. Cells give up their individual identity to cooperatively form something that has direction and purpose and will. Genes that were never active in the myxamoeba state are now fully responsive when the creature becomes a part of this giant organism. And genes formerly active when the creature was a shapeless bean bag are now silent. The change in the activation of genes is responsible for this unique collaboration of independent, unspecialized individuals.

It is a fragile cooperation, unfortunately. If the slug encounters a big pool of water, everybody disaggregates to go for a swim. Or if the slug encounters a juicy bunch of bacteria, all the members jump off the wagon to go for dinner. The slug is thus destroyed, and the slime molds resume their normal independent myxamoebae existence. However, a new fully cooperative slug can be reformed if that aggregating cyclic AMP signal is encountered.

Other aspects of the life cycle of the slime mold are fascinating. But the most salient point of interest here is its comment on life and death. No one would argue that the individual amoebae are alive. Given the right nutrients, they will split and divide for centuries. Yet when they form a slug, they create a new organism, also functionally alive. Is it just a matter of semantics to say that thousands of individual amoebae died so that one new slug could be created? The amoebae certainly have the ability to disassemble from the slug in the presence of water or food. When this happens, the slug is no longer in existence. It thus cannot reproduce, exhibits no animation, and has no sensitivity or feeling. Does the slug then die as the amoebae abandon it for more independent pastures? Do we really have an organism that can flit between life and death as easily as a butterfly flits from flower to flower?

The life cycle of the slime mold points out a fundamental flaw in defining biological demise in black-and-white terms. Death loses any sense of meaning when the transition between life forms is observed in this organism. Viability is a function of organization and, more to the point, gene activation. To the molecular biologist, the issue is not the assignment of arbitrary categories to a particular stage of development. The issue is simply observing which genes are turned on and off in any particular stage. This narrowing of life processes to the activation or inactivation of genes gives black-and-white definitions of death trouble.

Monolithic definitions of life and death would be further suspect if one found a fully functional organism that consisted only of genes. No cytoplasm. No nucleus. Nothing except nucleic acid, maybe a little protein, and a terrific ability to reproduce both. The status of the life of such an organism would be the most clearly assessed because the simplicity would give us the most unobstructed view of life. In His creativity, God has given us a window through which to observe life's processes in their most basic form. Such minuscule organisms actually exist. They are called *viruses.*

Even the Egyptians Got Colds

Viruses represent an extreme example of the problems of defining life and death. Viruses and their tiny cousins, viroids, are the smallest living things in the biological kingdom. Viruses have nucleic acid surrounded by protein in the same way that the creamy filling is surrounded by cake in a Twinkie; viroids consist only of nucleic acids, specifically RNA.

Most viruses are tiny biological terrorists. Typically, a virus hijacks a membrane protein to get inside a cell and then blusters its way into the nucleus. Once inside, like some computer hacker from hell, the virus completely reprograms the normal genetic machinery of its unsuspecting host. Instead of replicating the genes necessary for cellular survival, the virus forces the genes to replicate the viral nucleic acid according to strict specifications. This results in the creation of many new viruses and often results in the death of the host cell. And when the host dies, the newly completed viruses are spilled out into the environment and given a chance to infect other cells.

Viruses have forced us to reconsider notions about biological viability. Like some prewired robot, the life cycle of the virus is composed of chemical reactions. Those constituent chemicals can be placed into test tubes and various parts of the infection will take place in a totally predictable fashion. It is difficult to say when a virus dies because it is difficult to say if a virus is ever alive. Either it performs certain biochemical functions, or it does not.

Natural viruses are of such chemical symmetry that many of them can be crystallized, like a diamond or a ruby. In this hardened shape, they can exist for centuries. The oldest living virus was isolated from an ancient

Egyptian statue. When placed in a drop of water, the virus sprang to life just as it had thousands of years before. When the proper host was added to the tube, normal infection took place as if the virus had come from a twentieth century cough.

Biochemical Copying

Because of all these properties, scientists thought for a long time that viruses represented the simplest known biological creatures. However, this notion was superceded with the discovery of viroids, the champion minimalists in a room full of Philip Glasses. Viroids were first discovered as enemies of certain crops and remain mostly associated with plants. They have such exotic names as Coconut Cadang-Cadang Viroid and, no kidding, Tomato Planta Macho Viroid.

What are viroids? *Viroids* are single-stranded RNA molecules, like those messenger RNA molecules we talked about in the last chapter. Viroids took the traveling abilities of their messenger cousins a little too seriously. They have learned to exist outside the comfortable protection of a normal cell and, indeed, outside a living organism. This ability is extraordinary because RNA is usually quite susceptible to outside forces.

Viroids are not only hardy but also exceptionally small. In fact, they are the smallest living things known. Viroids are typically not more than three thousand nucleotides in length, about the size of three typical human genes. They possess no protein jacket like normal viruses. And they do not carry any protein baggage. They are simply tiny scraps of nucleic acid.

Viroids have one known function: they live to replicate themselves. That's it. Their entire genetic apparatus is geared toward performing a single reproductive act inside the nucleus of a cell. They seduce the resident RNA polymerase to attach to them and then replicate their genetic information.

The viroid has no intention of finding a ribosome so that its message can be translated. It creates no functional protein. It remains a nasty little stretch of nucleic acid forever and ever. The powers it exerts over the cell are so complete that in many instances, the cell dies. As a result, thousands of viroids are spilled into the environment. And the new viroids, like good parasites, look for new hosts to conquer.

Viruses and viroids represent the final chapter in an amazing reduction of complex organisms to their constituent parts. With these minimalists, their constituent parts are the only "living" portions of the organisms. Once we arrive at this level of simplicity, we see the problems with defining death in absolute terms more clearly. A virus lying dormant on a piece of Egyptian wood does not have feelings, and is inanimate, cannot reproduce. It is dead by the criteria mentioned previously, that is, until you put it into a drop of water. Then it becomes responsive, animated, and ferociously reproductive. Viroids as organisms totally consist of a single chemical response, their ability to be transcribed. A chemical response is incapable of feelings or animation. It reproduces because it can hijack a single set of someone else's enzymes. Asking if a viroid is alive under these conditions is like asking if a cola bottle is alive because it fizzes when you shake it.

Good Chemistry Between Us

And that is the point. From a molecular biologist's vantage, defining death by examining the ability to feel, reproduce, or become animated is futile. These three organisms—butterflies, slime molds, and viruses—demonstrate that viewing death as a monolithic difference between two absolute states is an absolute falsehood. When an organism collapses its cellular construction into a different pattern, death occurs only because the cells have been permanently rearranged. The creature may be able to organize and reorganize an infinite variety of times. Yet it does not swim between life and death with each oscillation. It changes its gene expression pattern, which in turn changes its shape and subsequent behavior. If the creature's biochemistry is simple enough, it behaves more like the physical compounds from which it is composed than a participant in one of two absolute categories. You can have organisms that still work after lying in a tomb for many thousands of years. You can also have creatures that remain perfectly functional by exploiting only one small portion of the genetic machinery of larger organisms.

We have seen that comparing death with life is useless until we have a definition of viability. We have also seen that the residual definitions of death are useless when applied to organisms. We are left with the tireless ebb and flow of certain biochemical reactions. If those reactions fail, the cell that carries them fails. If enough cells fail, the organism can no longer do what it was supposed to do. What does all this mean? It may mean that the best definition of life is the sum of certain critical biochemical processes, which in turn may mean that the cessation of simple biochemical reactions is the headwaters from which the concept of biological death flows.

High Interest Clones

Whenever I have been in a social situation and a discussion has started about the ethics of what I do in the laboratory, it's usually time to leave the party. The fear this technology engenders in people can be truly intense. I believe this reaction happens in part because genetics is so terribly personal. It also occurs because the genetic code has spent many centuries blissfully AWOL from scientific scrutiny. Now that genes are receiving attention as probably the most powerful set of chemicals in biological existence, unregulated freedom to tinker with them has come under question. And because this technology describes the very essence of biological life, many believe the technology should never be applied to humans.

One intimidating truth molecular biology has uncovered is that life can be transmitted to organisms like an Olympic runner passes a torch. It has also become evident that life, from organ to cell to gene, resides in constituent parts even in the absence of original organization. The definition of death, given these two facts, would be much clearer if we found that the buck stopped somewhere.

We already know that when genes are isolated, they can survive in solutions that would kill the hosts from which they were derived. We also know that these genes can be taken out of those solutions and stitched back into organisms. The transplanted genes work just fine. Thus, what renders an animal dysfunctional does not always render a gene dysfunctional. What does it take to render a gene dysfunctional? What do broken genes do to creatures that carry them? Will the study of these functions give us a clearer definition of the origin of biological death?

Land of the Three

To understand how genes are disrupted, we must review a little biology from the last chapter. You recall the meatball-Frisbeelike structure, the ribosome. The ribosome, that protein-synthesizing factory, finds a messenger RNA and immediately latches on to it. The first thing the ribosome does is to search along the message like a bloodhound on the trail of a fresh quarry. It is looking for something called a *start codon,* a very special region of the messenger composed of three nucleotides. Those three nucleotides tell the ribosome to quit searching like a bloodhound and begin making proteins. The ribosome dutifully obeys that message by opening up its protein-synthesizing factory and letting amino acids in. When it lets amino acids in, a functional protein is formed.

But which amino acids will it let in? Like some molecular ice cream shop, there are more than twenty amino acids to choose from. The only amino acids that get into a ribosome are those specified by the messenger. They must enter the ribosome one at a time and only as they are called up. It takes a combination of three nucleotides on the messenger RNA to code for a single amino acid, a triplicate message the ribosome can read very well. One group of three codes for one type of amino acid, another group of three codes for yet another amino acid, and so on. After it has found the start codon, the ribosome will look at the next three nucleotides on the messenger. Like a customer at that ice-cream shop, it will order the amino acid specified by those three nucleotides. (Actually, the amino acid is towed in by a tRNA molecule. The tRNA actually recognizes the codon in the messenger and, after recognition, dumps its amino acid in the ribosome.) After the amino acid arrives, the ribosome will look at the next three nucleotides and order yet another amino acid. When the second amino acid arrives, the ribosome will join it to the first one, thus creating a chain. After the joining, this protein factory will examine the next three nucleotides and order yet another amino acid. Eventually, an elongated protein is created.

The ribosome will continue this process until it gets to the molecular equivalent of a stop sign. Called a *stop codon,* this halt signal is also composed of a group of three nucleotides. When the ribosome encounters it, it stops ordering amino acids from the molecular menu. It then releases the newly made protein into the cytoplasm, and the message is discarded as waste.

This process of reading three nucleotides is effective in translating the desires of the chromosomes. If there are any defects in the genes that made the message, the message will deliver the error to the ribosome. The ribosome will faithfully translate the error into a protein, a protein rendered inoperable if the error is severe enough. Genetic mistakes, like a bad cold, are transmitted to the protein via the messenger because of a mutated gene.

How Geraldo Rivera Got That Way

A gene can be rendered dysfunctional in various ways. If a single nucleotide is removed, the entire message is shifted down one notch, and the subsequent protein is often crippled beyond the point of usefulness. If the message gets one more letter than it should, the entire message is again shifted by a letter, and the same effect on the protein is observed. These types of anomalies occur in nature and are called *frameshift mutations.* To see how these frameshift mutations exert their effects, consider the following letters:

T H E B O Y H A S O N E D O G

If you were to divide this collection of letters into groups of three, you'd get a message describing the number of canine pets a child has in his possession:

THE BOY HAS ONE DOG

But let's say you have a mutation in this sentence, a mutation that causes the letter *A* to be removed. If you then try to read the message in consecutive groups of three, you get an increasingly nonsensical message:

THE BOY HSO NED OG

This is an example of a frameshift mutation caused as the result of a single deletion.

Now let's say the message suffered a different mutation. Instead of losing one letter, it inadvertently obtained another letter, which we will call *Z*. Let's say the *Z* gets stuck between the *H* and *E* of the word THE in this sentence. If this message were then read in groups of three, you'd get the following cryptic sentence:

THZ EBO YHA SON EDO G

As a result of the addition of one letter, the entire message is garbled. Reading the letters of mutated sentences in consecutive groups of three causes severe translation problems in each case. The addition or subtraction of a single letter can cause the entire message to be misread.

The Code Less Traveled

Ribosomes suffer the same problems when they receive a mutationally altered message. Mutations generally occur in the DNA first, that is, the anomalous letter is added or subtracted in the gene. The RNA polymerase will faithfully copy the mutation right into the messenger RNA it is making. In the typical cell, the message is not sixteen or seventeen letters long; it is almost always one to two thousand letters long. When that mutated message gets to the ribosome, the protein that is made may be as garbled and as useless as the messages in the examples given here.

Sometimes pieces of DNA can break off chromosomes like runaway icebergs. These wandering chunks sometimes collide with other chromosomes and stick to them, destroying certain genetic functions and/or deregulating others. There are mutations that look like someone took a big bite out of the middle of a gene. Some of the most subtle mutations occur by switching one letter in the long nucleotide sequence of a particular gene. In the example below, the letter *T* has replaced the letter *B*.

THE TOY HAS ONE DOG

Because one letter is switched, the original meaning of the entire message is lost. In the ribosome, a switched letter results in the replacement of one amino acid for another. The outcome of this replacement can be that the original intent of the message is destroyed and the resultant protein rendered useless.

Single nucleotide switches can result in an even more serious problem than a misinterpretation of the signal. Single switches can turn a trio of nucleotides from specifying an amino acid to a trio that tells the ribosome to pack up and go home. These mutations specify a premature *stop codon,* that molecular stop sign we talked about earlier. When the ribosome encounters such a signal, it stops making whatever protein it was working on and abandons the messenger RNA. The mission of the gene is thus thwarted. We can place premature stop codons onto genes we isolate in the laboratory. When we stitch this gene back into a cell, we find that the original life of the gene has been stripped. It is rendered completely dysfunctional because the protein it is supposed to make is not synthesized.

Clone of Contention

The mutations that render a gene product dysfunctional may seem small and unimportant when examined in the laboratory. Nucleotide deletions, additions, or switches can cause incredible and sometimes fatal damage when examined in a human being, however. For example, the disease SCID (severe combined immune deficiency) completely wipes out an arm of the immune system. People with this disease must spend all of their lives in a germfree bubble because diseases that make most persons sick will kill them. You might think such an extraordinary disorder would have a wildly extraordinary cause. That turns out not to be the case. A single gene, *adenosine deaminase,* has one letter that is wrong in its code, a solitary nucleotide switch in the middle of the gene. As a result, the protein this gene makes is dysfunctional. This dysfunction causes certain cells of the immune system to behave awkwardly. The immune system is shut down, and the child is doomed to life in a bubble.

There are a number of examples of mutations in the body exerting a deleterious effect on the organisms that give them shelter. A lethal bladder cancer is caused in humans by the replacement of a single nucleotide in a gene called *ras*. This gene has thousands of letters in it. Nonetheless, if one nucleotide is off, an entire human is lost. There are numerous examples of human handicap, all caused by mutations occurring in single genes.

These examples of mutations demonstrate the ultimate power the genetic code exerts over biological creatures. Many parasitic organisms have recognized this power. Some organisms, like bacteria, have created proteins to attack foreign DNA as it enters the cytoplasm from a viral source. Some organisms, like certain viruses, create proteins that enter the nucleus and chop up the host DNA like men with butcher knives. The power of the genetic code is used as both defensive and offensive weaponry in cellular fights for survival.

Desert Storm in a Test Tube

Let's examine genes used as defensive weaponry. In the world of microbiology, tiny bacteria can be infected by even tinier viruses. These aren't viruses that would give the little bug a cold and go away in a few days. If the victim doesn't do anything to stop them, these viruses will kill the bacterium in about an hour. They do so by first injecting hostile DNA into the unsuspecting host. The foreign genes on this DNA code for proteins designed to hijack the genetic machinery of the bacterium and make more viruses. The bacterium dies, partially because it can't get to its own genetic apparatus and use it for normal function.

Obviously, the way to stop this infection is to destroy the injected DNA before it has a chance to set up shop. A bacterium relies on two mechanisms to do just that. The first mechanism involves the manufacture of a molecular hit man. Genes in most bacteria code for proteins called *restriction endonucleases*. The sole function of an endonuclease is to hunt for DNA, any DNA, and chop it up into little pieces. This immediately presents a problem because the host that made this little assassin also has DNA swimming in its cytoplasm. To distinguish foreign DNA from its own, the bacterium employs the second mechanism. It creates proteins with the ability to stamp its own DNA with a chemical marker that says, "I belong here. Do not chew me up." The endonuclease angel of death bypasses the host DNA and assaults foreign invaders. When a virus injects its DNA into a cell, the endonuclease finds the invader and chews it up.

Viruses, on the other hand, have a few tricks up their sleeves to combat host defenses. For example, a virus called *T4* also infects bacteria. Before its life cycle is out, it makes sure there are no witnesses to its crimes by producing enzymes that can chop up DNA. At a certain point in the life cycle of the cell, these enzymes are guided to the bacterial chromosome, which they promptly eat for dinner. The cell, now devoid of any genetic information, is rendered dysfunctional.

The conclusion is inescapable. Altering the structure of a gene, even if you mutate a tiny part of it, can completely destroy its function. A particular trait cannot continue to exist if the genes that manufacture it are destroyed. If the trait is necessary for functioning, the organism that houses the alteration will be crippled. We have finally found a biological entity that cannot be stripped down to independently functional components. Is it possible that this genetic inability to recover is a more valid definition of death? To answer this question, we must examine what happens when a scientist tries to take inanimate chemicals and create life from them.

The Life of the Party

Statistically speaking, the odds against existence are enormous. That says something either about existence or about statistics. The truths molecular biologists are excavating from the nucleus of the cell show that, if anything, the statistics were optimistic. Existence hangs on the coordination of thousands of correctly placed genes, each of which must have the sequence of even more thousands of correctly placed nucleotides. They must retain that order so that very low-energy chemical reactions can take place in a specific order. Life sinks its fragile claws into the genes, and the genes in turn attach their purposes to biochemical hooks.

Which is another way of saying that biological life is chemical reactions.

If that extremely controversial notion is true, we should be able to make some predictions about genes, prophecies that can be tested experimentally. For example, we should be able to reconstruct life by artificially reconstituting the specific chemical reactions. That is, we should be able to make nucleotides from their constituent elements. Then we should be able to build a functioning gene by placing those manufactured nucleotides in a certain order. The artificial genes should then spring to life when placed in naturally occurring microorganisms. We should also be able to mix and match our artificially synthesized nucleotides to make organisms that have never existed before. We should be able to take nucleotides from animals long extinct and wake the genes up by providing the proper chemical reactions, as long as the DNA is intact. The truth is that all of these possibilities have already been realized. We must deal with that fact when we consider what biological death really means.

Powder Power

Can a gene be artificially constructed from its elemental powders and exhibit all the properties of its natural counterpart? The answer is yes. And the process in recent years has become automated. You can buy prepackaged nucleotides from a wide variety of biotechnology companies. You can buy a speck of A or G or T or C for not much more than you would pay for a good pizza (with extra cheese). The nucleotides usually come in little brown bottles and are a lifeless crystalline white powder. If you don't want to buy these powders, they can be constructed by artificial synthesis in the chemical laboratory. They are manufactured from such things as carbon dioxide gas, found in most soft drinks, and ammonia, the active ingredient in Windex.

Once the purified nucleotides have been manufactured, the next task is to hook them together into a specific order. This involves placing the nucleotides through a series of specialized chemical reactions. The particular reactions are so well-defined that machines have been created to automate the process. These gadgets, termed *gene machines,* are marketed by biotechnology companies. They usually contain a small computer on board and places to insert your store-bought or home-brewed A, G, T, and C compounds.

To begin the reaction, you type into the computer the particular gene sequence you wish to create. The computer reads the instructions to the machine, and after a period of time, the gene appears in a drop of water at one end of the machine. These gene synthesizers cost about as much as a small mobile home, well within the range of most major universities and biotech companies. They have been used to manufacture many significant nucleotide sequences. In theory, they could be used to create any gene.

There are other ways of creating copies of the genes of interest. For example, you can isolate all the messenger RNA from a cell and put it in a test tube. You then have a complete collection of all the genes that were active at the time you isolated the RNA. Now you add an enzyme, *reverse transcriptase,* which chemically converts RNA back into DNA. When that enzyme is finished, you still have a complete record of the cell's activity. The only difference is that now the record looks like genes instead of messengers. You place these genes onto the molecular equivalent of tractor-trailer rigs so you can move them around. You can isolate, amplify, and otherwise manipulate any of these created genes at will.

Artificial synthesis can be started from the basics like ammonia and carbon dioxide or from the use of reverse transcriptase. These are only two of a growing number of ways to isolate and amplify genes. Any gene that has been cloned, sequenced, and placed back into an organism has gone through many rounds of artificial re-creation. But do these artificially constructed genes exhibit life?

The answer to that question is yes, depending on how you define life. Artificially constructed genes retain most of the functions of normal, unaltered genes. One could artificially construct an unregulated growth hormone, place it into a small mouse, and watch the mouse turn into a giant. By the same techniques, we can place part of the human immune system into mice. Or create tobacco plants that glow in the dark.

We have taken the genetic blueprints of viruses and made artificial ones from those instructions. The biological properties of a created virus are indistinguishable from those of the virus from which we stole the blueprints. We have exploited this technology to create new viruses that remarkably facilitate our cloning procedures. Biotech companies can greatly expand their businesses by providing researchers with improved forms of viral life. We use several forms of these artificial constructs in the everyday work of the laboratory so that we can be free to concentrate on ever-deeper questions about our own research. Such technology will eventually allow us to create more complicated entities than simple viruses.

Natural Synthetics

The experiments just described could have had a different ending. For example, we could have constructed the right nucleotides, placed them in the proper order, and done all our biochemical homework correctly. Then we could have asked our constructs to perform a normal biological function associated with life. If life required an additional magical ingredient, these constructs might have floated around lifeless in the test tube. Or equally lifeless when placed in a real organism. We might then conclude that we were missing some prime ingredient, some necessary otherness that powered the engines of existence. Biological death might then be defined as the absence of this otherness, even as biological life might be defined as its presence.

But that is not what we see. The bottom line of these experiments is that created genes, starting from gas and ammonia, behave no differently from genes found naturally. When we build organisms from these genes, the organisms spring to life and behave like their natural counterparts because the natural function of the genetic code is to facilitate miraculous but totally predictable biochemical reactions.

This thought is disturbing. Many people might be tempted to say that we cannot construct a single complex cell, let alone any larger organisms. It then follows that until we can, the jury is out as to whether death is the lack of certain biochemical reactions. Moreover, we cannot take a long dead organism, reapply certain biochemical reactions, and have it spring back to life. Or can we? We have seen that it is possible to jerk back into existence viruses that have been "dead" for many centuries. Is it possible that we can perform this same trick with parts of more complex organisms? Experiments on the horizon suggest that maybe we can.

You Really Can Blame It on Your Parents

The ancient Egyptians constructed mysterious tombs and had even more mysterious methods of embalming. Some mummies excavated from their tombs possess startlingly preserved soft tissues. One specimen dated around 300 B.C. contains a large organ package left in the chest cavity. The tissues of these organisms are so well preserved that they are amenable to scientific inquiry.

Molecular biologists can isolate the DNA from numerous tissues in Egyptian mummies. These DNA sequences can be placed on those genetic tractor-trailers we discussed earlier so that the genes can be manipulated and observed. We find that the genes, after all these years, are fully capable of chemical activity! Twentieth-century enzymes can duplicate the ancient molecules with every bit the fidelity of tissue isolated from my little finger. If engineered correctly, these genes can be sequenced, their ancient structures determined, and the RNAs from these genes reconstructed. If whole genes are present, they can be transcribed and translated into proteins that have not existed on this planet for thousands of years. Such analysis is being performed on tissues derived from frozen woolly mammoths, ancient sloths, and extinct horses. Scientists are studying the structure of ancient genes to understand their relationship to their modern-day equivalents.

Processes in our contemporary bodies are oddly quite similar to the rescue of the DNA from ancient Egyptian aristocracy. The sperm and the egg from which you were originally derived just as originally contained a rather mixed bag of genes, all contributed by people in your family long since dead.

Sound impossible?

Not really. Whether you are proud of this or not, you are the original product of half your mother's genetic information and half your father's. You've got half of each of them resident in the stuff that makes up your very being. Where did your parents get all this sophisticated genetic information? Your mother got half her coding sequences from your grandmother and half from your grandfather, which means your mother carried some of grandma's genes and some of grandpa's in her cells. In turn, your grandparents were made up of the genes from their parents. We have to examine only four generations before we are talking about gene sequence patterns in your body that existed in someone else's one hundred years ago.

The whole process repeats itself in a cycle that goes back to prehistory. Since your mom got some of those ancestral family genes from her parents, she was able to pass them on to you. Examining your chromosomes under the microscope is like strolling through a family historical museum of former noses, muscles, skin coloring, and body plans. These traits, all resident in you, were donated by people who have been dead for hundreds of years. And if you are yet childless, you are the latest contribution of the genetic efforts of those many, many people.

The point is quite simple. All life is the sum total of the chromosomes and genes within organisms. The genes are, in turn, the sum total of their biochemistries. These biochemistries are neither alive nor dead. Either they function in certain ways, or they do not. Such a conclusion is even more insulting and potentially more controversial than Darwin. He says that we share a common ancestry with apes.

The genes say we share a common ancestry with Windex.

Do Artificial Flowers Grow Best in Artificial Light?

Many people make the mistake of saying, "If I can't see it, it doesn't exist." They really mean that the light in the visible spectrum is not revealing anything to their eyes. Some wavelengths of energy are either too fast or too slow for our human eyes to register. Consider looking at a bowl of wine grapes, for example. A wavelength for purple bounces off those grapes, hits our eyes, and registers that color in our brains. There is a wavelength slightly faster than the color purple, however. We cannot see it with our human eyes; it registers to us as black. But the wavelength does exist, the color is called ultraviolet, and the light bulb that generates the wavelength is called a black light. Most people have seen the effects of a black light on certain paints. The eyes of some animals are capable of seeing in the ultraviolet range.

Concepts can be a lot like wavelengths of light. Some concepts absolutely exist but are very difficult to see at first because we do not have the cognitive eyes to register them. The subject of death is like that. It is a monolithic process only when you compare flying sea gulls in the sky with decaying sea gulls on the beach. But it is not some inexorable force that is either present or absent in all biological organisms. For most creatures, death is such a flexible process that it is optional. The only organisms exhibiting mandatory death are those that sexually reproduce. Since we sexually reproduce, death is mandatory for us. But we have a tendency to think that because death is compulsory for our biology, it is compulsory for every other biology as well. It's a lot like saying that because we cannot see an object, the object does not exist. In so doing, we misunderstand the composition of all the magical creations the Lord has given to this planet.

But even for us humans, unwillingly clothed in molecular caskets, life and death are not mutually exclusive processes. While we are alive, many parts of our bodies are calling it quits and shriveling away. More and more of our tissues die because of accumulating malfunctions as we grow older. That is why nobody dies of old age. We die because a critical function erodes to the point that it cannot support the body.

Yet when we die, much is still living. There are organs available for transplantation. There are cells and genes available for study. Your death will provide fuel for countless millions of organisms. Because they can coexist, death cannot be defined always in terms of the absence of life.

Such coexistence appears to do nothing to explain mortality, however. We have to do something with the dead seagulls on the beach, and eventually your own funeral with the children you leave behind. How can life and death be so dramatically different on the one hand, and yet be so thoroughly intertwined, on the other? We have stumbled upon a continuum. Life and death are the black-and-white terminals of a biological process colored in the middle with shades of gray.

Examination of the animal world bears witness to this continuum. Caterpillars are alive at one end of the wire, creep a little toward death in the chrysalis, and spring back to a lively start as the butterflies emerge. Slime molds traverse back and forth along this life/death continuum like trapeze artists. There is even room for the viruses and viroids in this continuum. To see their place, however, we are forced to investigate the composition of the wire that stretches between life and death.

The Life at the End of the Tunnel

The hint of death's composition comes from the fact that genes are the terminating point in the vast reduction of life. You can track life all the way from its organs to its genes and see an unbroken line to a particular trait. If that gene is destroyed, the trait it would normally donate to an animal goes with it. This dramatic association demonstrates that life in its final incarnation is composed basically of nucleotides. Get any of those nucleotides out of order and an organism can pay for the mistake with its life. The continuum, at least in part, must be composed of functional genes that confer information to the organism.

Another hint about the composition of the continuum comes from the fact that organisms can be built from their constituent genes in the laboratory. If the right nucleotides are placed into the proper order, a gene will be created. This manufactured gene will function as normally in cells and organisms as its natural counterpart. If the correct genes are clustered together, you can create an organism indistinguishable from one found in the wild. Thus, the same conclusion is derived from experiments executed on opposite ends of the question. If we tear down life to its basic components, we see that at the heart of those components lie genes. If we create those genes artificially, we can build simple life forms indistinguishable from creation. The same conclusion mentioned previously is similarly derived here, only this time with teeth. The wire strung between the life and death of any organism is made of genes.

The presence of genes is enough to assign structures to the life/death continuum. What pushes this presence toward one terminal or another totally depends on the genetic activity and biological coordination of the nucleic acids present. Egyptian mummies have lots of ancient genes inside them. So does the family history museum inside your cells. If a mummy's genes are to function, they have to be parceled and then awakened artificially. Your body doesn't have to be parceled to function, and you can rely on an alarm clock to awaken your nucleic acids every morning. The critical issue is whether the genes possess sufficient and coordinated activity to perform certain functions. Where we assign those combinations of functions to our continuum is a totally human judgment.

Circular Seasoning

It would be less disturbing if no organism was so simple that it was made only of a single set of chemical reactions. Viroids pose that intellectual dilemma. Viroids are just chemical reactions. They represent one small theme in a fugue built with such complexity that the simple tenacious biology of a viroid is almost absurd. Yet these tiny creatures exhibit life in its most pure and self-replicating form. They point out that active genes consist of chemical reactions. Since our life/death continuum consists of genes, the continuum is made up of this same biochemistry. The sum total of all biological life can be boiled down to the presence of certain chemical reactions. And the sum total of all biological death can be boiled down to their absence.

Defining life and death as two separate and mutually exclusive states is a mistake. It is like trying to get a garter snake to swallow its tail. The idea of a continuum may make many people feel uncomfortable in the biological arena. Yet any absolute concerning human life must be sprinkled over a legion of relative biological differences. We are made more uneasy because we live in an age where relativism is worshiped almost as much as money. And the opinion of the God we serve, who can perfectly traffic in absolutes, is ignored in favor of the comparative. On the canvas of human biological existence, few colors are more unwelcome in the religious ideological palette than gray.

However, neither scientists nor preachers created the biological kingdom. We are forced to come to grips with creation in whatever form it exists—even if certain definitions slip out of our hands and smear peanut butter over our intellectual laps. As we shall see, the idea of relative differences becomes very useful in placing certain biological experiments in perspective and in watching other perspectives make a mess in our hands and then quietly slither away.

Chapter 4

The Concepts of Conception

Introduction

I learned a great lesson in humility on the first day of second grade. It happened when I was wearing thick-lensed eyeglasses, the ones that come mounted as standard equipment on the eyes of every nerdy little seven-year-old. It all started because of a medical checkup (complete with an eye exam) I had undergone several weeks before. The extremely thorough appraisal included a psychological workup, consisting entirely of an IQ examination. The results of those tests came back several weeks later, the day before second grade was due to start. That afternoon, I managed to overhear my mother discussing the IQ evaluation with a friend on the phone and found that I had scored well in certain areas. I immediately stuffed that information into my little ego, quickly ran out the door, and promptly announced to all my friends how bright I thought I was. The rest of the afternoon, I planned how I would announce my discovery at school. Thinking that sitting in a classroom to learn was now beneath me, I mused that teaching might be more fun.

With this idea firmly planted, I eagerly entered the rear door of my second-grade class the next day. The teacher was at the front of the room, her back to me, shelving books. I exclaimed in a loud voice to her (and to anyone else) from the rear of the room: "I don't really need to be here. But I have come anyway to find out if there is anything I can do to help you. Thank you."

Proud of my generosity and magnanimous spirit, I took my seat. For about half a minute I looked around the room, savoring what my new role would be like. If I had known what was coming, however, I would have savored the next thirty seconds. It was the only calm moment I was to have in the second grade.

From the front of the room, an already stern-looking Ukrainian body slowly turned around from the bookshelf. At first she just stared at me. I don't mean she looked at me. I mean she stared a hole right through me. And slowly a fire began to burn in those eyes, a fire that could have barbecued an ulcer into Mother Teresa. She said, "Young man, you are going to need more than your glasses to help you at the finish of a typical day in my class."

She then turned around and resumed her book shelving.

Time seemed to freeze for an instant and then for the rest of the year. Thanks to my comment, I spent all of second grade in academic solitary confinement. It was practically grounds for expulsion if I so much as flinched without asking. Or so I thought. Nothing that I did was good enough to warrant faint praise. By the middle of the year, I was beginning to think I was mentally retarded; by its end, I was convinced of it.

I didn't realize at the time that I was learning an important, if humbling, principle of life that year: we sometimes have to pay for the consequences of our reactions to certain discoveries. Some of those consequences are good. Some of them aren't, as my experience in the second grade can attest. The payment can cause us to overdraw our moral bank accounts. It happens because sometimes our brains become equipped with knowledge that our hearts are unequipped to handle.

The social implications of investigative research can sometimes carry the same kind of payment principle as my second-grade experience. This is not due to the nature of the investigation; there is nothing intrinsically wrong with finding out your IQ scores. The problem is with the nature of the investigator and with the population that decides what to do with new knowledge. Science could care less about ethical decisions. Even though it can discover the great technologies of the universe, science is powerless to tell us what to do with them. Humans alone have the ability to make ethical decisions. We have to superimpose what our minds discover onto our emotional and spiritual abilities to implement what we find. It is a simple problem of vision. Many times science, with deep intellectual eyes unaided by thick-lensed glasses, has seen farther than our hearts can look.

These principles come into sharp focus when we begin to superimpose scientific insight onto moral questions and religious beliefs. To me, it is quite clear that God is crazy about human beings. It is also quite clear to me that God does not feel the same capacity for human tissue. He probably does not grieve, even as we do not grieve, over the loss of fingernails we just trimmed. He probably does not weep over the disposable products of a haircut. In turn, we do not pray that an amputated arm accept Jesus into its, what, knuckles? before it dies. The conclusion is that there is a distinction between human beings and human tissues. Worth is infused into one and not the other. What causes this difference in attitude? What is the difference between human life and human tissue?

In this chapter we are going to learn some basic molecular processes governing conception. This knowledge is critical if we are to make a rational decision about human life and nonlife. Along the way, it is probably going to be tempting to place a value judgment on a biochemical process. That's not unusual; almost every generation throughout history has done it at one time or another. Where we draw the line and where someone else draws the line is tricky if the issues are clear-cut. And biological issues are seldom clear-cut. The problem with molecular biology's making certain processes so clear is that it makes the moral issues so fuzzy. In the laboratory, it is one thing to call something alive. It may be a whole other thing to call it human.

Multiplication by Division: The Strange Math of Human Reproduction

Humans have always been fascinated with that most ancient of blendings, the ability of two genders to produce a third human being. Although science has been able to take much confusion out of the process, it has been unsuccessful in removing even a little bit of the mystery. Among primitive peoples, reproduction was so inexplicable that the process itself was worshiped. They created those embarrassing statues of gods that looked like sexual organs and worshiped them. Some advocated the male role as the exclusive reproductive power, some the female role. In a habit that has contemporary corollaries, the tribes that advocated these differences often used them as an excuse to have a war. Conflict meant never having to say you're sorry.

The ancient Greeks, of course, were among the first to season the reproductive miracle with biological reasoning. Aristotle wrote a treatise that eventually got him nicknamed the Founder of Embryology. He described somewhat accurately the development of the chick and other embryos. He even made a stab at human reproduction by saying that babies formed as a mixture of menstrual blood and semen. His prescription for getting pregnant probably didn't go over very well. Perhaps a better name for him would have been the Semi-Founder of Embryology.

Eventually, it all got straightened out by technological advances that revealed the facts. There were such things as sperm and eggs. Those cells held heritable genetic data; babies were a Cuisinart blending of that information. Further, babies had certain traits because chromosomes, contributed by both mom and dad, were jumbled together. Finally, the chromosomes contained DNA, which could be divided into discrete traits we have termed genes. Thus, the mixture of genes determined the gender, size, strength, intelligence—in short, the life—of a particular human.

We now know that this mixture of genes has a very defined way of reproducing itself. As far as humans are concerned, these reproduction modes fall into one of two similar-sounding categories, meiosis and mitosis. Even though the names sound alike, they are quite different processes. But they are quite necessary for the survival of the species. Without the first process, we could not grow into babies. Without the second, we could not grow into adults.

Split and Polish

Let's first consider *mitosis,* which has been briefly mentioned (see fig. 4.1). Mitosis begins with the forty-six chromosomes making doubles of themselves. The chromosomes do not look like the familiar little X's that you may have seen in pictures, however. During this replication stage, the chromosomes look more like hopelessly entangled spaghetti noodles in a big round pot. Fortunately for us, the intertwining is not disorganized.

The entangled chromosomes soon slither away from each other and condense to form those little X's. Seen under the microscope, each chromosome has a partner, which means that the cell possesses ninety-two chromosomes (forty-six plus forty-six). These chromosomes line up along the center of the cell like a column of little soldiers in the middle of a battlefield. Soon a cellular fault line forms down the middle of the cell as if the chromosomal army were experiencing a molecular earthquake. At this point, the cell is splitting in two, separating the chromosomes into two distinct camps. Shepherded by stringy little pulleys called *spindle fibers,* forty-six chromosomes will go into one half of the dividing cell, and forty-six will go into the other. In this fashion, the cell clones itself, making a fairly exact twin in only a few hours.

Mitosis is the reproductive method of choice in human tissues. The process is employed by 99.9999 percent of the cells in the body. Only a comparatively few cells undergo the second type of cellular division, *meiosis.* The cells made meiotically are useless except under very specialized circumstances. Even so, we absolutely depend on this process for survival into the next generation. The cells that undergo meiosis are eggs and sperm, also called *germ cells.* Without such a division process, babies would never be formed.

Survival of the Lightest

Meiosis begins in a manner very similar to that of mitosis (see fig. 4.2). All the chromosomes make duplicates of themselves. These twins pair up

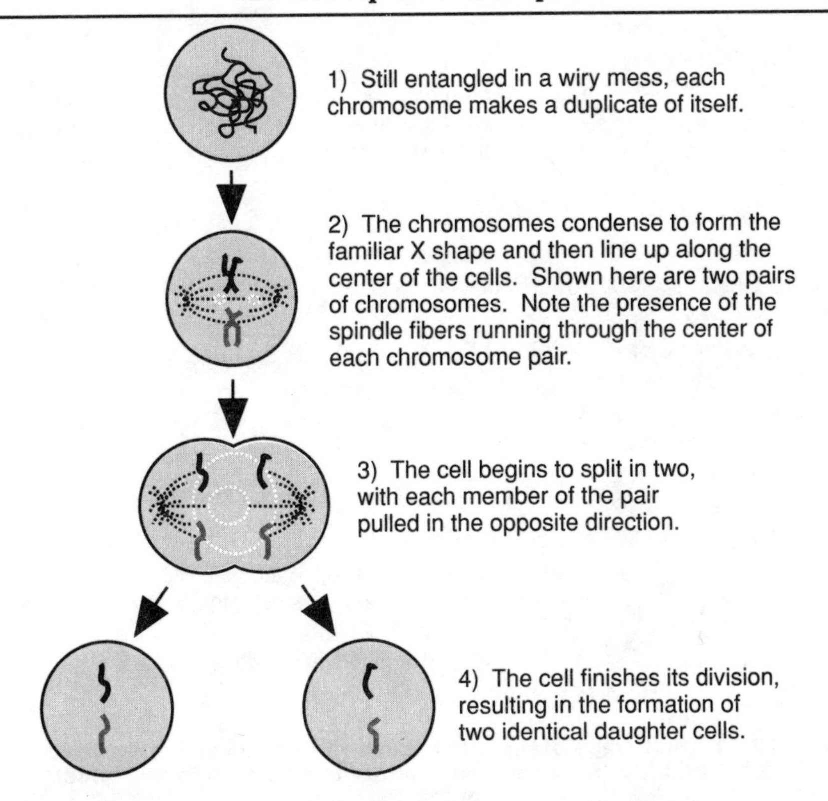

***Fig. 4.1.* Mitosis results in two identical daughter cells being formed from a single cell.**

with each other, like I had to do in second grade when my notorious teacher made me hold hands with Bobby Smith, whom I couldn't stand. Chromosome one sidles right up next to the replica of the chromosome it just made. Chromosome two pairs with the replica of chromosome two and so on. Each twin is now called a *homologous chromatid* because scientists like fancy names. The total count of the chromatids is ninety-two, just like the mitotic division.

At this point in meiosis, a weird thing happens. In an orderly fashion, the chromosomes break open in certain places. They form *chiasmas,* which is Greek for "we don't really understand this process." The chromosomes, once broken, literally exchange chunks of DNA with each other like they were at a swap meet. They swap by funny rules, however. A chromosome will exchange pieces only with itself or with the twin it made in the previous replication step. Sometimes only the tips of the chromosomes will exchange. Sometimes only small pieces in the middle will exchange. Sometimes huge chunks will swap with each other (see fig. 4.3).

The end result of this biochemical bartering is dramatic; parts of the

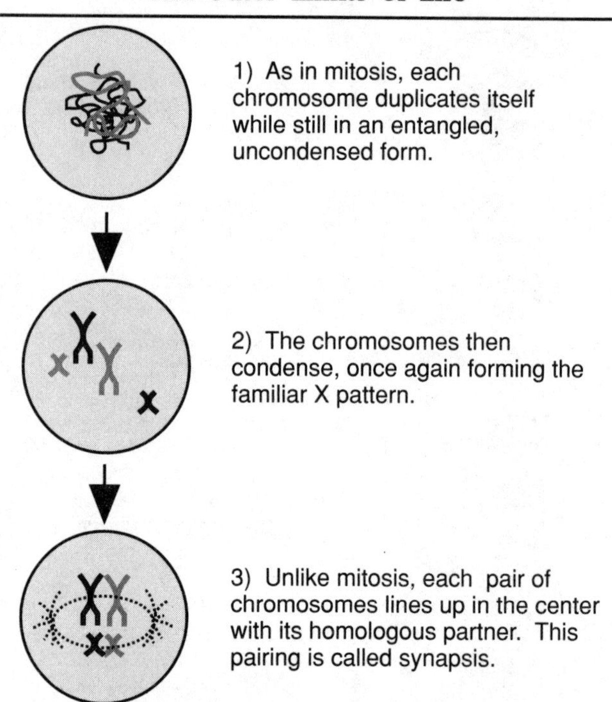

***Fig. 4.2.* The first stage of meiosis begins with chromosomal duplication. Each pair finds its homologous partner and lines up in the center of the cell.**

chromosome are now made up of the original sequence, and parts of the chromosome are made from the results of the swap. The chromosomes, while still fully intact, are now a patchwork of gene sequences. This exchange confers genetic variety on the part of the cells undergoing meiosis and confusion on the part of the scientist studying the process.

After the exchange has occurred, the cell containing these ninety-two chromosomes settles down and gets ready to divide. The chromosomes, already lined up in the middle, are pulled to opposite ends of the cell by spindle fibers. Just as in mitosis, forty-six chromosomes move to one end of the cell; forty-six move to the other. While these chromosomes are moving, that familiar fault line forms in the center of the cell again, and the cell splits in half. Each of the two cells contains a somewhat jumbled collection of forty-six chromosomes. This is called the first *meiotic division.*

A second meiotic division follows to accomplish a chromosomal weight-loss program. It "loses weight" because the two cells that contain forty-six chromosomes undergo an additional cleavage without first undergoing a replication step. When the second split occurs, each cell will

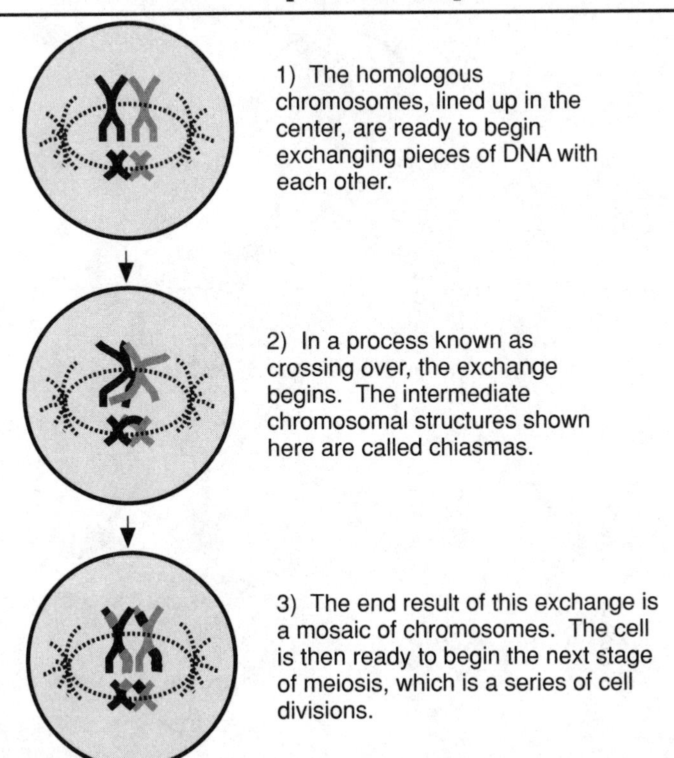

***Fig. 4.3.* The next stage of meiosis results in a genetic mosaic. With homologous chromosomes paired in the middle, the strands of DNA exchange segments with each other.**

be rationed only twenty-three chromosomes. The second meiotic division is thus unlike any other duplicating process known to human cells (see fig. 4.4).

It looks fairly normal under the microscope, however. The forty-six again line up in the center of the cellular parade field. The familiar fault line occurs, and the cell begins to divide. With the assistance of the spindle fibers, twenty-three chromosomes are pulled into one cell, and twenty-three chromosomes are pulled into the other. Four cells are made (two from each product of the first meiotic division), each with half the genetic information of a normal cell.

More Than Half Off

This process accomplishes two essential biological objectives. The first is variety. If you have any nontwin brothers or sisters, take out their photographs. It is easy to see that though you share the same parents and

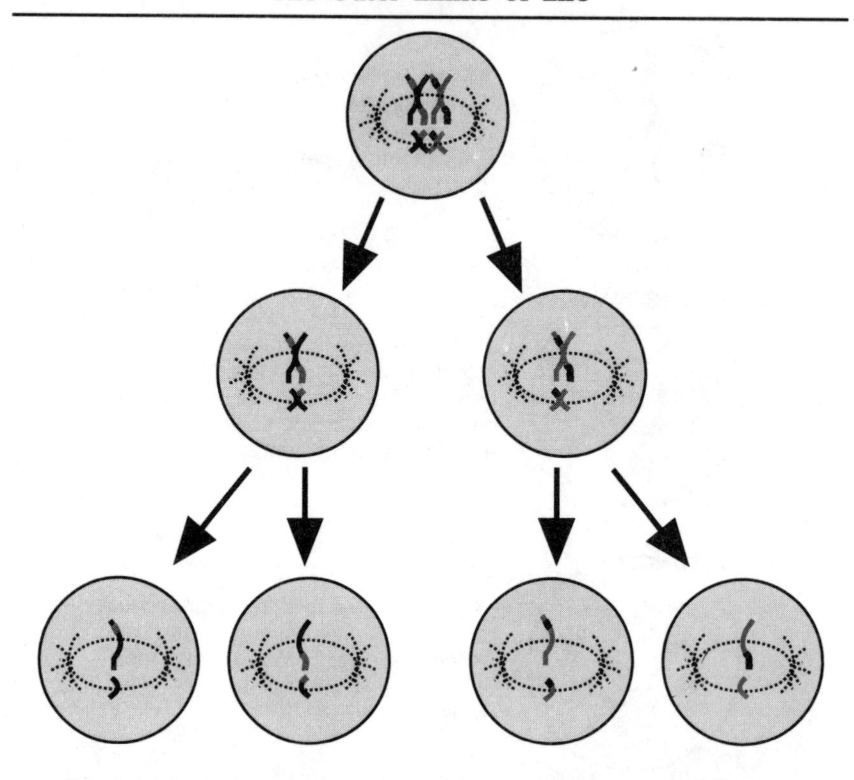

Fig. 4.4. **The last stage of meiosis results in a series of reduction divisions. The first division is very similar to mitosis. The second round of division is very unlike mitosis, however. Half the number of chromosomes exist in the daughter cells when this round is finished.**

may have some similar features, you are not identical. That is due to the genetic swap meets in your parents' genitals prior to the acts of intercourse that produced you and your siblings. One sperm got your father's nose in the swap meet. Another sperm got Aunt Hazel's nose. One egg got your mother's eye color; another got Uncle Harry's. When the sperm that got Aunt Hazel's nose met up with an egg that had your mother's eyes, the result was a child born with both those traits. Thus, you and your siblings are genetic mosaics of all the swap meets the egg and the sperm encountered prior to the first meiotic division. This exchange ensures that we are not uniform and genetically inflexible clones of each other.

The second biological objective is reduction. Mitosis produces cells with forty-six chromosomes. That is fine if you want continuously reproducing adult tissue. It is biologically useless, however, for creating babies with genetic variety. You need two cells that each possess exactly half the required genetic information. That way, when the cells join, they will manufacture one complete human being. Meiosis allows both conditions to be met. Our very survival depends on the ability of this process to accomplish those goals.

What does the study of meiosis and mitosis have to do with deciding what is a human being and what is a human tissue? Human cells that undergo mitosis, like those in your big toe, divide like the tiny denizens inhabiting a drop of high-school biology pond water. We don't call either big toe cells or pond-water monsters human beings. Cells that undergo only mitosis can never become little babies because they make too many chromosomes every time they split. Human cells that undergo meiosis, like sperm and eggs, have a much better shot at making human babies. But gurgling little infants can never be formed if the only thing that happens to sperm and eggs is that they lose a little chromosomal weight. The question of human tissues versus human beings thus bounces like a Ping-Pong ball between the biochemical engines that drive mitosis and meiosis, which is unsatisfying. At this point, the clarification of these biochemical issues does not result in the clarification of their moral constituency.

Return to Gender

The difficulty ancient humans encountered in understanding the biology of reproduction rivals the difficulty modern ones find in communicating reproductive information to their kids. It wasn't until the advent of the microscope that we learned anything more than the Semi-founder of Embryology had written. With the help of the microscope, it became evident that the cells we now know as sperm and eggs were somehow involved in reproduction. A common belief was that the really important cell was the sperm, most probably because men controlled the sciences, not to mention the money and the armies. The idea was that sperm carried in its little head a prefabricated miniature adult. When that sperm encountered a warm uterus, the baby was dumped inside it, triggering a nine-month enlargement cycle. The only thing the egg provided was a restaurant where the sperm could eat during the pregnancy.

We now know that the egg and the sperm have fairly equal input into the genetics of the developing baby. The only controlling aspect left to the sperm is the choice of gender; the male decides if the baby is going to be a boy or a girl. A male child doesn't come magically equipped with these decision-making abilities at puberty. He has to first grow the sperm cells in the replicative process of meiosis. This event occurs at puberty; the sperm cells in the young boy's testicles undergo meiosis and become mature. By placing certain biochemical properties on preexisting tissues, the male becomes reproductively competent.

Meiosis does not automatically confer on cells a Bachelor of Protectable Humanity degree, however. And so our original question remains about what actually does. Is the nature of the constituents inside the individual sperm and egg cells given the gift of sentience? Does God look down on sperm and egg cells and say, "You exist; therefore, you are human"? To understand the nature of the question, we must look at the anatomy and development of the individual players. We shall begin by studying these human sperm cells, both their individual structure and their genetic origins.

Letting It Go to Your Head

The structure of normal human sperm, like most dinosaurs, consists of a head and a tail. The head contains all the information necessary to penetrate an egg and herald the genetic good news that a baby is on its way. The tail contains all the structures necessary to get that information to the egg.

The primary parts of the head consist of a nucleus and a structure termed an *acrosome* (see fig. 4.5). The nucleus of human sperm contains all the vital genetic information necessary to create exactly half a human being. The nucleus contains only twenty-three human chromosomes, often called the *n number* by scientists who have a hard time with math. In addition to creating 0.5 baby, the nucleus of a sperm decides junior's total gender. If the nucleus contains a Y chromosome, the baby will be a boy. If the nucleus contains an X chromosome, the baby will be a girl. Because the egg can only contribute an X chromosome, the egg plays no part. Male embryos are XY and female embryos are XX. Equal numbers of both male- and female-producing sperm are in the semen of male human beings.

The sperm is practically on automatic pilot after ejaculation has occurred. It is already genetically primed to travel to its destination and needs no further transcriptional information to perform the single task of finding an egg and joining with it. Thus, mature sperm nuclei do not synthesize messenger RNA. Moreover, except for the rear of the cell, the nucleus contains almost no nuclear pores. A *nuclear pore* is a hole in the nucleus that permits communication with the outside world. The lack of these perforations may mean that there is no need for such communication while the sperm is in transit. Human sperm cells may act something like preprogrammed robots, needing only a boarding pass to a warm uterus to fulfill their genetic destinies.

The female reproductive tract exerts certain biochemical effects on the sperm head once it's inside. In the process of *capacitation,* the female

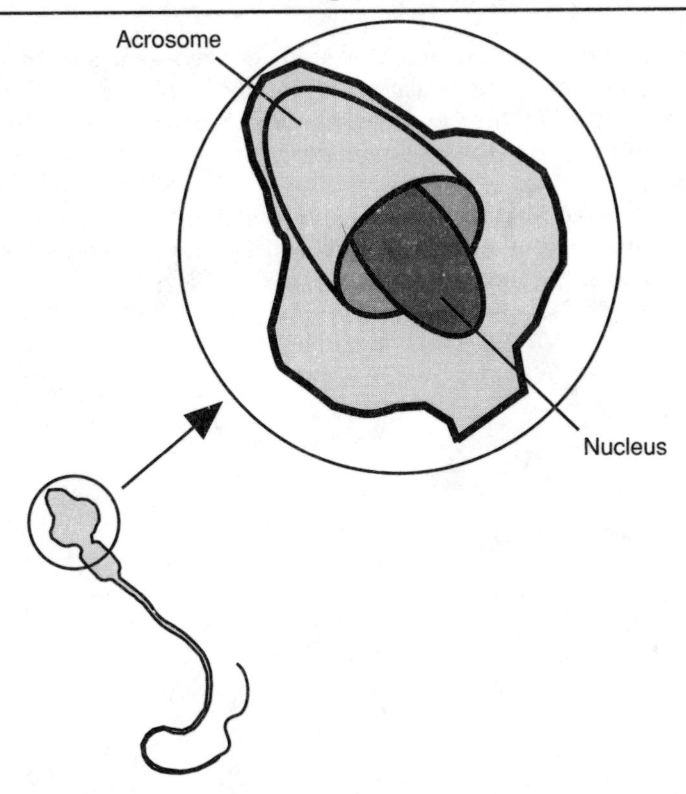

Fig. 4.5. **Human sperm, head region magnified (internal structures exposed)**

causes changes to occur in the membranes and chemical constituency of the sperm head, specifically the acrosome. Without these changes, the sperm would never be able to penetrate a human egg. Thus, the female confers on the sperm the final ability to perform their functions. The sperm actually have to remain in the female tract several hours before capacitation is over and fertilization can proceed.

Fertilization does proceed, with sperm conducting the information in its nucleus through the uterus. However, all that information would be useless without a means to drill into an egg after the sperm got there. Now that the sperm has been prepared, a tiny organelle on the tip of the head is ready to perform its function. This tiny organelle, an acrosome, is the chemical equivalent of a battering ram. Under the microscope, the acrosome looks like a thick inverted drinking cup, with the sperm nucleus tucked inside it.

The acrosome is filled with enough biological explosives to make a munitions expert envious. Inside this little bag are enzymes that can digest an enormous number of proteins. There are also enzymes that can digest lots of sugars. There is an enzyme so specialized—and so ferocious—that the sperm keeps a molecular muzzle on the enzyme's mouth until it's ready to use it. All of these enzymes have one purpose: to be released from the acrosome and detonate on the surface of the egg when a sperm encounters it.

The task of delivering this dangerous information falls to the second part of the human sperm, the tail. The tail consists of two parts: the mid-piece and, confusingly, the tail (see fig. 4.6).

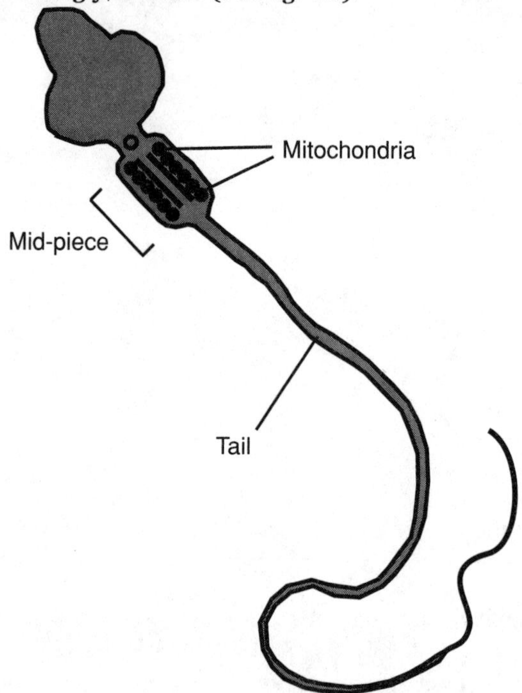

***Fig. 4.6.* The tail region of human sperm**

The Sperm of the Moment

The *mid-piece* contains the fuel that powers the motion of the sperm. In the mid-piece are tiny batteries called *mitochondria,* which look a little bit like allergy capsules under a microscope. These little organelles pump out energy by digesting the degraded products of a whole grocery store of biochemicals. These biochemicals include fructose, the sugar found in honey. They are all stocked like rocket fuel in preparation for the long journey. Mitochondria also exist in cells other than sperm. But the ones in sperm contain a protein that specially toughens these batteries for the trip. This protein exists in no other cell in the body. But then, no other cell in the body was designed with the single function of leaving.

The other section of the tail is called the *tail* by scientists. The most important internal structure in the tail is an *axoneme,* a long series of filaments that look a lot like helical bundles of uncooked spaghetti noodles. These bundles are made up of interconnecting proteins that can slide and ratchet past each other. This ability to slide and ratchet produces a frenetic motion that catapults the sperm past the cervix and onward into the uterus during intercourse.

And a curious motion it is. You may have seen films of sperm motion, wiggling and writhing. The motion is actually initiated at the base of the tail and is propagated in a waving circular motion throughout the cell. This causes the sperm head to spin 180° every time a wave of motion occurs in the tail. This whirling, twisting motion does its job, however. Sperm can creep along an unobstructed surface at about five inches every hour.

What has just been described is the mature, postpuberty, I'm-ready-to-make-a-baby sperm cell. For thirteen previous years, however, the human male just described was incapable of fathering anything. Did mature sperm cells all of a sudden confer reproductive sexuality on the owner? Were there cells in some kind of coma for better than a decade inside the little boy? To answer these and other questions, we must understand the mechanisms of gender differentiation. And to accomplish that, we must peer into the mysterious world of life in the first month after conception.

The Molecular Biology of Macho

The determination to be a girl or boy depends upon the presence of an X or a Y chromosome in the sperm. The genetic gender decision is thus made at conception. The genes that will eventually shepherd the operation initially leave the embryo alone, which permits genderless growth for the first seven weeks of life. The fetus develops what appears to be a genderless sex organ so uncommitted that scientists have named it the *indifferent gonad* (see fig. 4.7). This is probably the only time in our lives when we do not have some interest in sex.

Around the end of the seventh week, this asexual biological Garden of Eden ends. The fate of the indifferent gonad is to become a sexual organ. The fetus changes that organ into whatever gender is specified by the genes. Once that organ is established, it will secrete the hormones that will convert the rest of the fetus into its genetically assigned role. So the whole point of sex determination is to establish specific organs and then let those organs turn the embryo into its proper gender.

To comprehend how this strategy applies to the manufacture of masculinity, we have to digress for a minute. A curious fact about human developmental biology is a slap in the face to insecure males. Embryological and genetic data demonstrate that all humans are biologically hard-wired

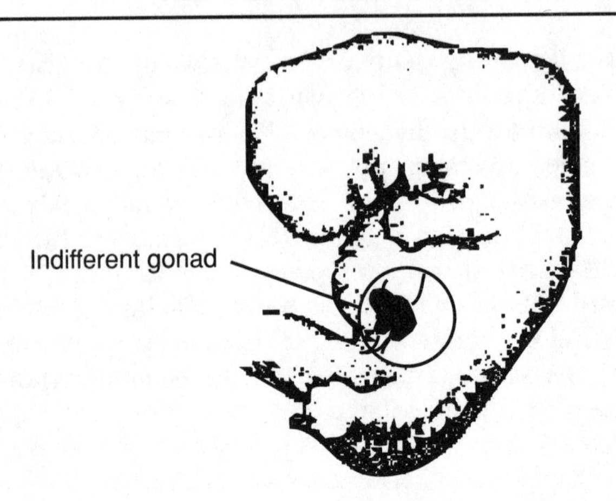

***Fig. 4.7.* The indifferent gonad as viewed in a seven-week-old embryo**

—courage, guys—to become women. If an embryo never sees any of the so-called male or female hormones, ovaries will start to develop in the fetus anyway. Animal experiments show that if you remove the ovaries at an early stage, the fetus will form a uterus, a vagina, and other female sex organs. If part of the fetal male apparatus fails to shut down certain cells, the biological male will also contain a uterus, a vagina, ovaries, and PMS. There has to be active intervention from an outside source to stop this process from occurring. In the case of establishing the masculine gender, the Y chromosome executes the police action to turn a human fetus into a male. The curious thing about masculinization is that most of it is controlled by a single master gene. In humans, this gene is called SRY. The rest of the Y chromosome can be discarded and, as long as this gene is available, a testis will form. This was illustrated in a dramatic way: the mouse equivalent of the SRY gene was implanted into a female mouse embryo, and the little embryo changed genders. It grew up to be a male mouse.

The developing genitals react very quickly to the presence of the SRY gene product (also known as *testis determining factor*). The indifferent gonad, no longer indifferent, responds to the factor, and its insides slowly but irrevocably turn into the familiar testicle. And immediately this organ has a job to do; many cells could still convert the fetus into a female. This new testicle secretes two hormonal weapons: *Mullerian inhibiting substance (MIS)* and *testosterone*. MIS is a biochemical whose sole job is to hunt down the cells that could make the female reproductive apparatus and functionally kill them. The other hormone, testosterone, is busy proselytizing other cells to the masculine doctrine, affecting the development of the brain, stimulating cells to make the prostate gland, seminiferous tubules (more on those later), and so on. Eventually, the fetus turns into a male, with external genitalia visible by the end of the first trimester.

So have any sperm cells formed yet? We certainly have a testicle now, an organ that secretes masculinity like Donald Trump used to secrete wealth. The answer to the question about the existence of sperm is yes, they are in the fetus. Sort of. Beginning at the third week of life, fully a month before the MIS terrorist strike, certain cells form in a structure called a *yolk sac*. The yolk sac is a big beach ball–like structure attached to the belly of the tiny embryo (see fig. 4.8). Inside that yolk sac, cells named *gonocytes* start getting restless. These gonocytes, which will be future sperm cells in the male, have not yet undergone meiosis, so they contain all forty-six chromosomes. They crawl out of the yolk sac and slowly move toward the indifferent gonad. At this stage, these cells don't look anything like human sperm cells. They look more like those amoebae we talked about earlier. It takes them almost two weeks to arrive at the indifferent gonad. But they could have taken even more time than that because they will not be recruited to do much of anything for almost thirteen years. By that time, the baby has been delivered, and a screaming little boy has magically taken the place of a screaming little infant.

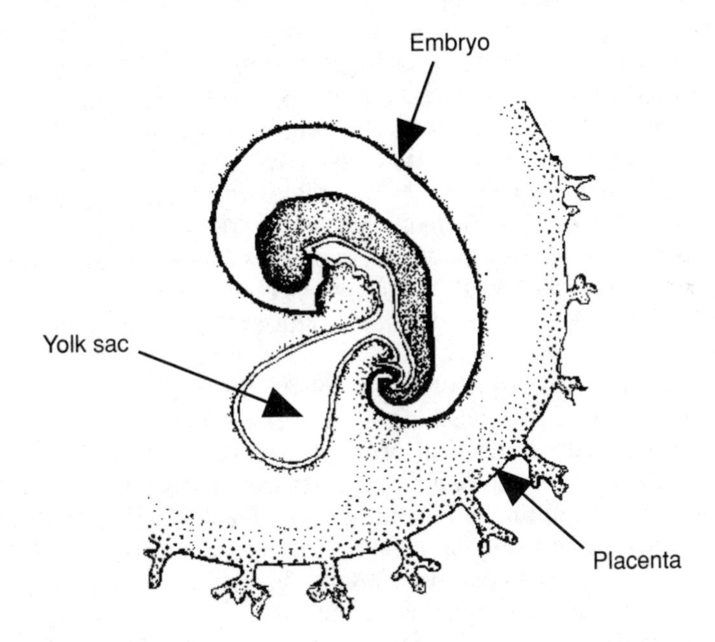

Fig. 4.8. **The yolk sac in a developing embryo (shown here at three weeks of age)**

Puberty Ruins Everything, Part I

The next focus in terms of reproduction potential shifts to the little boy's brain, specifically the pituitary gland. This gland can be thought of as the arsonist that will start the fires of puberty in the little boy's body. One of its first tasks is to tear down several structures in the immature testes and replace them with new ones. This replacement wakes up those immature sperm cells that arrived at the gonads years before in the womb. These cells undergo a round of mitosis, an action that immediately caused us scientists to think up a new name for these cells. We came up with the term *primary spermatocytes* to describe the fact that these cells are no longer taking an extended vacation. We also came up with a new name to describe where these cells are now living as the result of pituitary action: the *seminiferous tubules,* which cradle the resting primary spermatocytes in a many folded, long tube.

The spermatocyte isn't actually in the seminiferous tubules. It's really in the arms of a very fussy nurse cell, or Sertoli cell, lurking deep within the folds of the seminiferous tubules. The Sertoli cell is the molecular tutor of immature sperm cells. The youthful spermatocyte will develop literally within her cytoplasmic arms. Under her tutelage, an immature spermatocyte will grow up to be, in about sixty days, a mature sperm cell (see fig. 4.9).

A resting spermatocyte, enfolded by the Sertoli cell, does not look anything like a mature sperm cell. It looks more like a small round grape. And the grape possesses forty-six chromosomes, having undergone only one round of mitosis since its wake-up call. This number is useless if the goal is to help make babies. The genes in the spermatocyte therefore decide to lose some chromosomal weight. They direct the cell to submit to the process of meiosis. When the process is over, the spermatocyte still looks like a grape, but it now possesses only twenty-three chromosomes. This event, of course, incited us scientists to give this same cell a new name: a *spermatid.*

Still within the arms of the shepherding Sertoli, the little spermatid begins maturing. The grape narrows in shape at one end until it assumes the almost arrowhead form of mature sperm. It gradually acquires the chemically caustic acrosome next to its nucleus. It gets a mid-piece, complete with its amazing little batteries. And finally it gets its tail, possessing the spaghettilike axoneme running all the way through its body. By the time the Sertoli cell is finished with it, the sperm cell is almost fully functional and ready to begin its egg hunt.

I say almost because even though the sperm looks finished, it is incapable of wiggling or fertilizing an egg. There is one final maturation step before the sperm can be ejected, a step the Sertoli cell cannot supervise. So the Sertoli cell releases its charge, and the sperm migrates through the

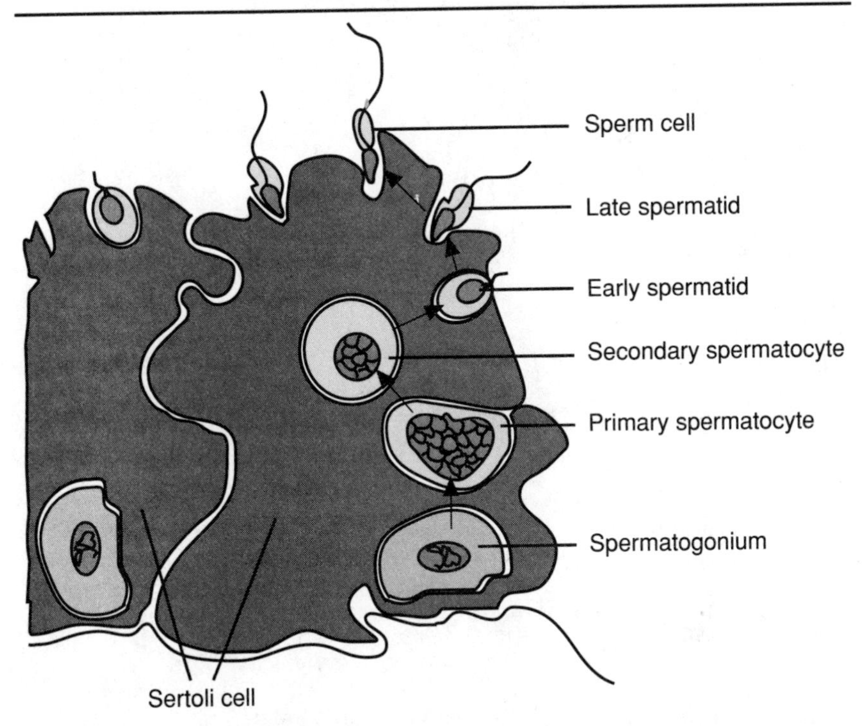

***Fig. 4.9.* Development of human sperm from spermatogonium to semimature sperm cell. All stages of development take place, assembly line style, in the Sertoli cell.**

maze of the seminiferous tubules (see fig. 4.10). It eventually finds its way into the *epididymis,* where it remains for several days. This structure functions as the finishing school for the developing sperm cell. The acrosome is remodeled, and the structure of the membranes surrounding both the head and the nucleus is altered. Changes occur in the number and types of small biochemicals within the cytoplasm of the cell. As a result of these cumulative changes, the sperm obtains the ability to move. When capacitation occurs, the sperm will be able to fertilize any human egg with which it comes in contact.

Although the process of manufacturing a human sperm is a miracle of design and development, it is by no means free of mistakes. Humankind holds the disconcerting record of having the greatest percentage of malformed, immotile and anomalous sperm in the animal kingdom. Some sperm have two heads instead of one, others refuse to wiggle, and still others have oddly shaped heads. This is true even of men of proved fertility. The explanation may lie in the fact that we are the only creatures in nature to routinely wear clothing. The temperature of the testicles may be raised above physiological tolerance because of the insulating power

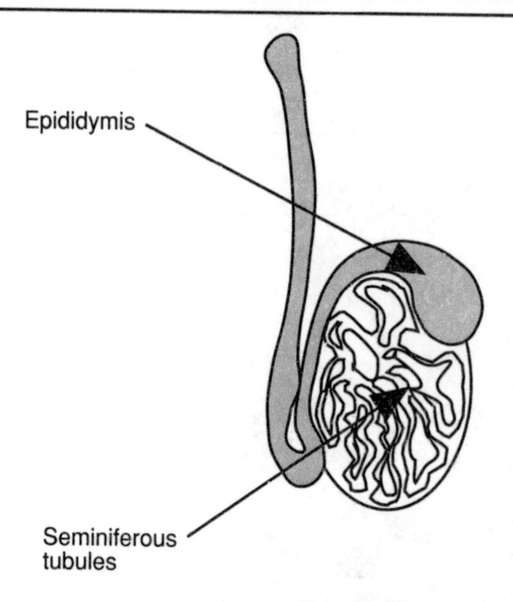

Fig. 4.10. **Internal architecture of the human testicle. The developing sperm find their way from the seminiferous tubules to the epididymis where final maturation takes place.**

of underwear and pants. Indeed, when other animals are similarly artificially insulated, the percentage of malformed sperm also rises.

Fissures of Men

Whether a man has healthy sperm or not, what does the manufacture of those cells have to do with deciding whether something is human life or tissue? We generally do not seem concerned about the fate of individual sperm, probably because they have no capacity to become human beings if left by themselves. There are no national prosperm special interest groups, and we have no crushing moral dilemma in extinguishing the biological potential of sperm. Nature apparently has little problem with it, either. Out of 400 million possible candidates in a single ejaculation, only one will be used in fertilization. The rest will die.

Moreover, the manufacture of sperm is not a uniquely human process. The basis for much research into humans has come from studying lower life forms like mice and sea urchins. The most startling observation about such research is the similarity of the processes from one organism to another. Sea urchin sperm have wiggling tails powered by mitochondria, and mice possess motherly Sertoli cells. Thus, there is no claim to a unique personal characteristic because human males can make sperm in their testicles.

Nonetheless, some features of sperm cells, though shared in common with other animals, are unlike those of any other cells in the human complement. Biologically, they possess only half the DNA of normal cells. Their genes have been shuffled and rearranged in a fashion unlike any other cell type or other sperm. This incredible individuality confers on the sperm cell a very special characteristic: the potential for the development of a human being. Sperm cells may not be people, but some persons say they are clearly more than human tissue.

If human sperm cells are not human beings, what about their corollary cells in the female? Are human eggs, taken by themselves, any closer to being human persons than sperm? Or your big toe? What is so special about their development that a partnership with wiggling sperm can produce football players and scientists and musicians? To get better insight into the nature of human life and human tissue, we must understand a little developmental biology of human ovaries and their contents, the subject of the next section.

Slaving Over a Hot Ovum

Historically, the female experience has always incited men's contradictory attitudes. By possessing the vault that houses the preborn, women have wielded the power of life and death over the entire race. Through the years, many societies have relegated their women to the humiliating position of property, perhaps as a reaction to their power.

The ideas concerning their biological role in reproduction have been equally schizophrenic. The idea had been historically developed that sperm carried the prefabbed kid around in its little head; the female was needed only as a food-stocked rental trailer for a while. Virtually at the same time, the opposite idea was being promoted that the egg held the miniature human. Semen was vitally involved; it had some kind of life-conferring property that magically kicked the kid into active mode during intercourse. Many felt that the wiggling sperm cells within the precious semen had nothing to do with reproduction. The good stuff was in the fluid, not in the cells that inhabited it.

It wasn't until the early part of this century that the proper role of the egg was established. The surprising discovery was that this reproductive oval of the ages worked in humans a lot like it worked in other creatures. Humans lay eggs like many animals. We just don't do it in the same environment. Turtles manufacture eggs and lay them in the sand. Fish lay their eggs in the water. Birds lay eggs in aerial twig-lined baskets. Human females, however, lay eggs inside their own bodies. Ever distrustful of the thermal instability of our environment, we execute our pregnancies internally. Our biology reacts to this suspicion by allowing us to usually produce only one offspring a year. This internality has produced a curious addiction. Our offspring fully depend on parental nurturing skills for a length of time that exceeds any in the animal kingdom. These days, that length of time can be measured in decades.

How does this egg come to exist in the human female? Do eggs exist only when puberty occurs? What does all this mean at the level of the genes? We know that the gender is determined at conception. If the sperm contributes an X chromosome, the developing baby will be a female. Exactly how that little girl acquires the ability to return the reproductive favor given by her parents is described in this section.

Premarital X

The female fetus begins her preegg manufacture just as early as the male begins making his presperm, about the third week postconception. They use exactly the same cell to elaborate their gender. The amoebalike cells that crawled out of the yolk sac in genetic males also crawl out of the yolk sac in genetic females. Like the males, they have forty-six chromosomes and are called by the same familiar name of gonocyte. Even their target is the same; these gonocytes crawl out of the yolk sac with the specific intention of finding the indifferent gonad and lying down beside it.

Only during the eighth week of life do we observe a distinctly feminine difference in the embryos with female genetics. Without the Y chromosome, the Mullerian inhibiting substance does not exist. There is little testosterone. With none of these substances to alter development, the indifferent gonad in the genetic female commits to developing an ovary. The creation of this organ will greatly affect the fate of the gonocytes, which set up shop near the indifferent gonad and were present at the eighth week.

The first thing that happens to these gonocytes is that they divide like crazy. This division is pure mitosis, and as a result, the cells retain their forty-six chromosomes. The second thing that happens is that we scientists give these dividing cells the new name of *oogonia* so we can increase the number of English words beginning with two vowels. By the third month postconception, these oogonia have divided so much that they exist like clusters of grapes surrounded by primitive ovarian tissue. Every cluster arises as the direct result of the division of a single oogonial cell. Each cluster in turn is surrounded by an audience of flat cells. Sent to these clusters by the primitive ovary, the flat cells will eventually exert a dramatic biochemical effect on the growing oogonia (see fig. 4.11).

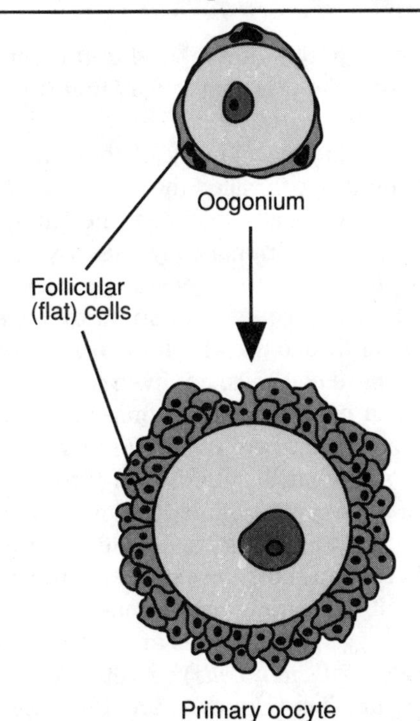

Fig. 4.11. **The oogonium differentiates into the primary oocyte. The follicular (flat) cells will eventually secrete oocyte maturation inhibitor.**

Most of these clusters continue dividing by mitosis, happily maintaining forty-six chromosomes for many months in the womb. This is an extremely active period of replication in the little ovary; the number of oogonia will swell to over 7 million by the fifth month. After a while, many of these cells will either change into something else (more on that in a minute) or start to die away. Most of the oogonia will be dead by the seventh month. If nothing else happened to some of the oogonia before that time, human females would be born reproductively sterile. The reason for the massive increase and sudden loss is not yet known.

Something else eventually happens to a fraction of the oogonia, an event that saves every generation of humans from certain extinction. This something involves the oogonia's chromosome number. If the oogonia continued to divide only by mitosis to maintain their forty-six chromosomes, they would be reproductively useless. A few cells in some clusters seem to share the same genetic insight. Some of these cells quit dividing by mitosis. Instead, they start the initiating stages for a round of meiosis. This division, as you recall, will eventually lead to the halving of their number of chromosomes. These committed oogonia also distinguish themselves from their surroundings by putting on a lot of cellular weight;

these cells look bigger than the others under the microscope. This change, of course, told us scientists that it was time to give these new cells another name: *primary oocytes.*

As the primary oocyte gets bigger, more of those attending flat cells surround it. The entire entourage is called the *primordial follicle.* Even as the number of oogonia dramatically declines, the number of primary oocytes with their attendant cells dramatically rises. By the time the baby girl is born, she will have 700,000 to 2 million primary oocytes in her ovaries. It is fortunate that she arrives with so many because she stops making them as soon as birth occurs. The human female will never get another chance to make more egg cells, not even at puberty. Eggs, unlike most cells in the body, can never replace themselves.

These precious eggs are taken care of in an intriguing fashion in the developing baby. While an infant, her body drugs the oocytes with the molecular equivalent of a sleeping potion. This isn't what the eggs are designed to have happen to them. Eggs are genetically programmed to carry out the meiosis division process they started in the first place. Meiosis is stopped dead in its tracks, however, because of the actions of the flat cells surrounding the eggs like groupies around rock stars.

The flat cells (also called *follicular cells*) secrete a hormone known as *oocyte maturation inhibitor.* This hormone functions like a sleeping pill whose effect over the eggs lasts for at least ten years. The primary oocytes cannot complete meiotic division as a consequence of the hormone's action. Only with the kiss of puberty will these cells wake up and finish the meiotic process initially begun before birth. Two inhibiting phenomena are seen in the genetic control of these cells: a limit is placed on the number of eggs that can spring forth from a mature woman's womb, and little girls are not allowed to produce them.

Puberty Ruins Everything, Part II

So these eggs, these primary oocytes, are allowed to sleep in their warm ovarian castle for more than ten years. After that time, some decisions start to be made in the head of the little girl that will awaken these eggs to a brand-new environment. Way up in the pituitary gland, that air traffic controller of incoming adulthood, some interesting transactions have taken place. The genes of the pituitary have been in chemical negotiations with the genes of another part of the brain, the *hypothalamus.* As a result, the hypothalamus sends the pituitary an emissary by the name of *gonadotropin-releasing factor,* mercifully shortened to GnRF.

As part of the deal, GnRF instructs the genes in the pituitary gland to release the hormones LH and FSH (which stand for the fact that scientists can abbreviate things every bit as competently as the Defense Department can). Both hormones are released into the bloodstream and journey immediately to the undeveloped reproductive apparatus of the owner. They aren't exactly handsome princes, but they very effectively wake up genetic machinery that has been asleep since the little girl was born.

FSH finds an oocyte surrounded by those sleeping-pill-secreting flat cells, termed a follicle as you recall. FSH wakes up this sleepy follicle—both flat cells and the egg—by stimulating their genes to produce certain biochemicals. This is not surprising, since FSH stands for *follicle stimulating hormone*. LH also participates in this maturation process. Among the biochemicals stimulated are the familiar estrogens, hormones that can make long-distance phone calls to the pituitary gland. This communication between brain and reproductive organs is the opening theme in the familiar symphony of reproductive maturity. The cells that remained in a state of inactivity for so long one by one begin to fulfill their reproductive potential.

What does this mean to our little primary oocyte, now yawning and stretching its reproductive genes for the first time in years? Remember that the egg had just started the process of meiosis before it was put to sleep. Now that it is awakened, it finishes the meiotic division it had started a decade before. It has that normal swap meet we mentioned earlier, with different homologous chromosomes exchanging chunks of genetic information with each other. The cells undergo the first meiotic division, which means that an egg produces two cells of forty-six chromosomes each.

The odd thing about these two cells is that, unlike what happens in normal meiosis, they are unequally shaped (see fig. 4.12). One cell basically gets a mouthful of discarded nuclear junk along with its chromosomes. It isn't even called a cell; it's a *first polar body*. The other cell gets all the goods along with its chromosomes. It is a healthy cell and is called a *secondary oocyte*. This first meiotic division, creating both the polar body and the oocyte, occurs shortly before the process of ovulation in the sexually mature female.

All the future genetic action takes place in the secondary oocyte. Remember that meiosis consists of two cellular divisions, called the first and second meiotic divisions. The secondary oocyte has started only one of these divisions. It would like to finish it and then start the second meiotic division if it could; after all, it has been waiting to complete this process for years. In fact, you can take a human egg out of an ovary and put it into a petri dish. The spindle formation, which marks the beginning of the second division, will occur shortly after the cells are placed in culture.

Back in the body, however, it is a different story. Just as the egg starts to make preparations for that second division, it is unceremoniously evicted from the protective warmth of the ovary, and some flat cells are thrown out with it. This is the process of *ovulation*, where the egg leaves

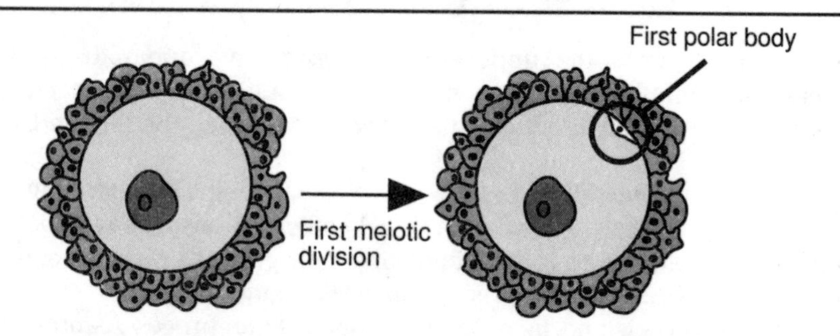

Fig. 4.12. **The first meiotic division of the primary oocyte results in two cells, a normal-sized oocyte and the tiny first polar body.**

the ovary and is coaxed to the outside world by the seductive wavings of Fallopian tubes.

This ejected egg hasn't completed its second round of meiotic division. And there is not much time for it. The little egg will not survive outside the ovary for more than twenty-four hours unless it has a close encounter with a sperm. If the sperm never reaches the egg, the egg will degenerate and die without undergoing its second meiotic division. A cell, which has spent years in preparation for a single event, is washed away in a crimson tide of menstrual waste if fertilization does not occur.

The human female will have many chances to fertilize her eggs. Only one egg is awakened and ejected at a time in the normal menstrual cycle. All the rest are still asleep. If she is still menstruating by the time she is fifty years of age, the eggs she will be releasing will have been asleep for over half a century! The sexually mature woman thus has an amazing reproductive capacity. If all the eggs she carries were to somehow become fertilized, the average human female could produce over forty thousand human beings. In her lifetime, she will introduce to the world over five hundred eggs as candidate offspring.

Not an Egg to Stand On

What does this prolific manufacture of human eggs have to do with deciding whether something is human life or simply tissue? The problems with human eggs, perhaps not surprisingly, are similar to those presented with human sperm. Oocytes alone do not have a sufficient molecular capacity to become human beings all by themselves. We do not save viable ova from menstrual waste for use in other contexts. We have no crushing moral dilemma in extinguishing their biological potential. As with sperm, nature apparently has little problem with it, either. A woman throughout her reproductive years will menstrually donate less than 1 percent of all the eggs available for fertilization. Of those women who will have children, less than 1 percent of the eggs that are donated will go on to create babies. At menopause, those eggs still within the ovaries will die.

Moreover, the manufacture of eggs, like the manufacture of sperm, is not a uniquely human process. The basis for much of the research into humans has come from studying lower life forms like mice and sea urchins. Sea urchin eggs undergo meiosis, and mice possess FSH. Thus, there is no claim to a unique personal characteristic because human females produce eggs in their ovaries.

Nonetheless, some features of eggs, though shared in common with other animals, are unlike those of any other cells in the human complement. Biologically, they possess only half the DNA of normal cells. Their genes have been shuffled and rearranged in a fashion unlike any other cell type or other eggs. This incredible individuality confers on the egg a very special characteristic: the potential for the development of a human being. Human eggs may not be people, but some persons say that eggs are clearly more than human tissue.

It Is More Blessed to Conceive

So how do a sperm and an egg join together to produce a human being? Perhaps an explanation of this process will give us a better clue about the meaning of human life versus human tissue. The joining of sperm and egg is the culmination of years of biochemical development, hibernation, and timing. It is one of God's greatest miracles. Most research examining this miracle has centered on the fertilization processes of the common laboratory mouse, and I will call upon this data frequently in this section. But as we have seen, most animals that sexually reproduce have very similar cellular mechanisms. Fertilization can be thought of as a giant singular theme, with only small variations among animals of different species.

During intercourse, the sperm and the egg must first find each other. Sperm swim up the female reproductive tract assisted by their own frenetic wiggling and the welcoming muscular contractions of the uterus. The sperm usually meets the egg in the Fallopian tube, a considerable distance away from the site of the deposit of either participant.

And what exactly does the sperm see when it happens upon a human egg bobbing up and down in the Fallopian tube? It sees bigness. A typical human sperm next to an egg has the same relative size difference of a paper clip next to a basketball. It sees some attending, secondary cells around the egg, often called *cumulus cells* (also termed the *corona radiata*). The sperm also sees a thick outer barrier called the *zona pel-*

lucida that surrounds the surface of the egg much like our atmosphere surrounds the earth. Underneath this barrier is the membrane of the egg itself (see fig. 4.13). The task of the sperm is to get through the barrier of the zona pellucida, fuse with the membrane of the egg, and dump its nucleus into the interior. When that task is finished, the sperm will die.

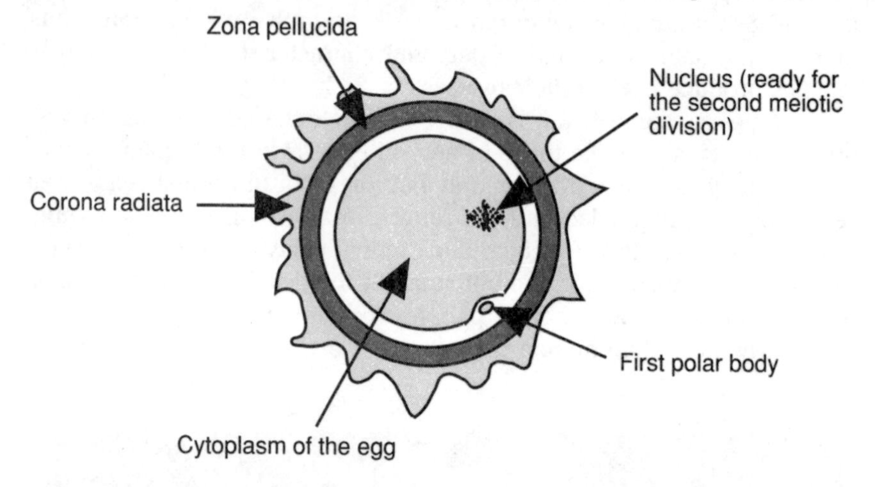

***Fig. 4.13.* A mature egg awaiting fertilization. The corona radiata is a layer of follicle cells generally expelled with the egg during ovulation. Note that the nucleus of the egg is poised to undergo its second meiotic division. It will do so immediately if it is fertilized.**

Although often idealized in books and movies, the meeting of sperm and egg is not an intimate introduction by two shy cellular lovers. Instead, many thousands of sperm converge on a single hapless egg in a frenzied and confusing mass of excited biology. Initially, many sperm can fasten to the surface of the egg. This first tethering is a relatively loose, nonspecific association. In embryological circles the complex scientific term of *attachment* is used.

Once attachment is completed, progressive deepening of the biochemical relationship occurs. Both the sperm and the egg have proteins on the surface of their cells that are made to dock with each other. In the mouse, the docking protein on the zona pellucida of the egg is called *ZP3*. ZP3 is designed to specifically interact with proteins on the surface of sperm cells. And dock they will; an estimate is that the egg hangs out almost 1 billion copies of this molecular red carpet on its surface. That is why so many sperm can initially attach to the egg. In the mouse system, as many as fifteen hundred have been found to attach to ZP3 proteins on a single egg.

ZP3 provides another important function on the egg's surface. It acts like an immigration border patrol whose job is to prevent sperm from other species of animals from fertilizing the human egg. ZP3, because it is a protein and thus looks like a pretzel, has a very specific shape, one that only proteins on sperm of the same species can recognize. It is impossible for any other species of sperm to fertilize this egg because there is nothing to grab onto. ZP3 provides an instant barrier against making unnatural hybrids. Recently, however, researchers have found that sperm from certain animals can fertilize eggs from other species if the zona pellucida containing the ZP3 or ZP3-like protein is first removed.

No kidding.

Let's ignore the implications of that research for a while. Back at the Fallopian tube, we see that hundreds of suitors are vying for the molecular hand of the single egg. What mechanism decides who the lucky fertilizing sperm is going to be? How does the sperm penetrate the outer layers of the egg and reach its cytoplasm? The answer to these questions involves looking more closely at ZP3 on the egg and also involves examining that bag of chemical munitions in the acrosome of the sperm.

ZP3 has sugar molecules on its surface. Sperm love sugar and will not bind to ZP3 unless the sugar molecules are present. In experiments where the sugar molecules are removed from the surface of an otherwise intact egg, sperm cells ignore the egg. In other experiments where only the protein with its sugars is supplied in a salt solution, the sperm bind to the molecules in the salt solution. You sweet-toothed readers out there have something in common with human sperm.

These sugar molecules are vital in the next phase of fertilization, the so-called acrosome reaction. Since the sperm nucleus has to unite with the egg nucleus, the point is to get one sperm bound to the ZP3 protein to detonate the contents of its acrosome on the surface of the egg. This will burn a hole into the zona pellucida large enough for the sperm to slip through.

How is this munitions factory stimulated to dump its contents to the surface of the zona pellucida at the right time? When the sperm binds to those sugars, the sperm gains the ability to drink calcium, the same molecule your mother said was in milk and so you better drink it if you want to grow up to be as big and strong as your dad. Perhaps she should have included the words *as virile*. When that calcium comes rushing in, the acrosomal membrane changes position. Initially tucked inside the sperm head, it now moves outward and fuses with the head's outermost membrane. This fusion allows the contents of the acrosome to be vomited onto the surface of the zona pellucida (fig. 4.14).

Once the caustic contents of the acrosome hit the surface, a large hole forms. The acrosome contains enzymes that can eat all kinds of molecules, from proteins to sugars to fats. Since most of these susceptible molecules exist on the surface of the egg, the egg is unable to resist the effects of the acrosome. After interacting with another zona protein called

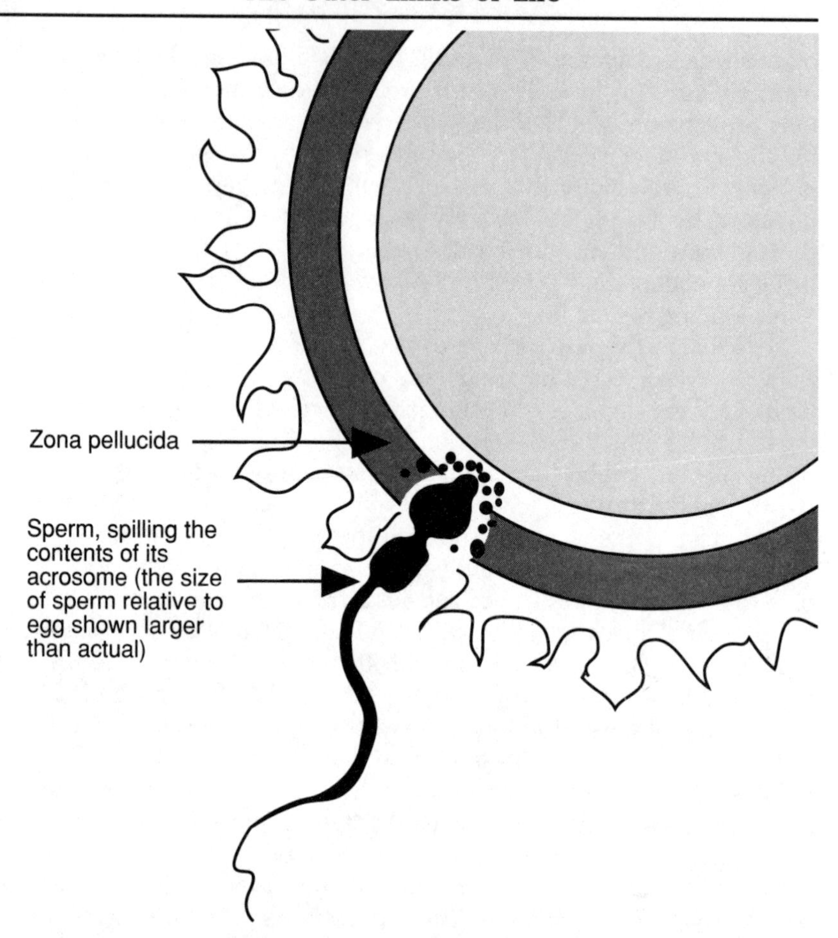

Fig. 4.14. **Human sperm penetrates the zona pellucida by releasing the contents of its acrosome. Once the enzymes of the acrosome have chewed away the zona pellucida, the sperm's membrane fuses with the egg's membrane, and the egg is fertilized.**

ZP2, the wiggling sperm is free to happily burrow its way deep into the egg, assisted by the residual effects of acrosomal release.

Not So Cold Fusion

The next step is to fuse with the membrane of the egg and deliver the nucleus of the sperm to the cytoplasm of the egg. To understand how this fusion occurs, we need to understand a little bit about membranes. To do that, we have to talk about, of all things, Mazola cooking oil.

Have you ever taken a teaspoon of Mazola cooking oil and placed it in a glass of water? You can still see the oil as droplets on the surface of the water. If you swirl the water really fast, you can see that these droplets will break apart into even tinier droplets. As the water begins to settle, you might notice that these droplets can fuse back together. You can even get them to form one giant droplet in the center of the glass. This can happen because the droplets obviously possess the ability to fuse with each other and they have absolutely no ability to fuse with the water around them. The droplets can be separate entities, or they can unite to form a single entity.

Under certain conditions, the same is true of human cells. Most human cells can be thought of as a drop of water inside a ball of grease. The grease of these cells is a very defined membrane made of oils. Proteins bob up and down in this membrane like buoys in an ocean. The internal drop of water is the cytoplasm containing the nucleus and the resident essential molecules. When human cells are placed under specific conditions, many of their oily membranes will do exactly what the Mazola did in your glass of water. That is, they will fuse together and share their internal contents as a result of the fusion.

This is precisely what happens in fertilization. The head of the sperm, no longer containing the nasty acrosome, is made only of its oily membrane, its tail, and its precious cargo of nucleus-enclosed chromosomes. The acrosomal reaction has rendered what is left of the sperm's membranes susceptible to fusion with other membranes. This acrosomal reaction has also rendered the tail, while still very much attached to the head, permanently motionless. Underneath the protective zona pellucida, the only barrier between the egg's nucleus and the sperm cell is a similar oily membrane. After the sperm has burrowed its way past the zona pellucida, the membrane of the head finds the membrane of the egg and re-creates the Mazola oil experiment. That is, the head partially fuses with the egg and in so doing dumps the nucleus into the egg's cytoplasm. The egg, sensing that it has been penetrated, grabs what is left of the sperm and swallows it, tail and all. In a move rather reminiscent of the praying mantis, the egg eats the sperm for dinner—all except for the precious nucleus, which floats unmolested in the egg's cytoplasm. But more on that later.

Nuclear Waste

The egg, with its captive male nucleus firmly secure, must now execute a couple of very quick maneuvers. There are hundreds, perhaps thousands, of sperm still knocking on her molecular doorstep. If any of them fuse with her membrane the way the first one did, a condition called *polyspermy* will be created. This condition, fertilization of a single egg by more than one sperm, is lethal and must be avoided. Within three seconds of penetration by the sperm, the egg deploys three first-strike chemical weapons.

1. *Electrical reaction.* As soon as penetration has been perceived, an electrical change occurs across the egg's membrane; it undergoes *depolarization.* This depolarization is due to a change in the kinds of molecules the egg is allowed to swallow. This lightning-fast change in drinking content alters several properties of the membrane. One property is the ability of a second sperm—or any other sperm—to penetrate the membrane of the egg underneath the zona pellucida. Some researchers say this property is either nonexistent or irrelevant in mammalian reproduction. Therefore, the electrical reaction in humans after fertilization is an area of controversy.
2. *Hardening of the zona pellucida.* The egg is not content to protect her outer skin. She also goes after any sperm that may have performed an acrosomal dump and started burrowing their way inward. The egg sends out a signal that turns the fairly porous zona pellucida into quick drying cement. Sperm cells that were already chewing their way into the egg are stopped dead in their tracks. They are trapped in what turns out to be a semen graveyard; the trapped sperm will die and be resorbed by the body.
3. *Cortical granule discharge.* Finally, the egg wishes to protect herself from the remotest chance of other suitors courting her (and they are legion) in her most intimate parlors. For this purpose the egg possesses thousands of little jars of chemicals called *cortical granules* beneath the surface of her cytoplasmic membrane. Inside these cortical granules are enzymes with the ability to chew up sugars. As soon as the sperm is swallowed by the egg, those cortical granule jars fuse with the egg membrane and dump their contents outside the cell. When those sugar-eating enzymes are released, they find every ZP3 they can and chomp on those sugars. Remember that sperm have a sweet tooth; the sugar must be present so that the sperm can bind. If the sugars are destroyed, no more sperm can bind to the egg, which is the point. Called the *cortical reaction,* this wave of instant sugar eating causes the egg to swell slightly outward. You can tell when an egg has been fertilized by looking for a slight change in size. When all these reactions have occurred, an egg has been fertilized by a single sperm cell and is fully incapable of being pierced by any other sperm.

State of the Union

So now what? The sperm's nucleus is floating in the egg's watery prison. For a minute the egg ignores that nucleus and embarks on a task it has wanted to complete for decades: it finishes its meiotic division. You recall that when the woman who just got fertilized was still a baby inside her own mother's uterus, she was producing eggs. You recall that the eggs she made tried to reduce their chromosomal number from forty-six to twenty-three by undergoing a round of meiosis. The rest of the baby, obviously chemically annoyed, stopped the process about midway through the second division.

Once this little girl became a woman, the body still seemed annoyed at the process. As soon as the egg was awakened and started to resume meiosis, it was immediately kicked out of the ovary. The egg, right up to the time the sperm bores its way into her cellular domain, has forty-six chromosomes. She must get the roster trimmed to twenty-three and get there fast to make a baby with her intruder. It's a do-it-at-the-last-possible-minute mechanism. The sperm that swirl around her underwent their reduction to twenty-three chromosomes almost sixty days before they entered the uterus.

So the egg has to catch up. As soon as it senses that it has been penetrated, the forty-six chromosomes in the egg line up hurriedly right down the middle of the cell. They then split into equal chromosomal halves. That is practically the only thing equal about it. The egg once again relegates twenty-three of those chromosomes to a wimpy little vesicle, really a junk heap, on one side of its sphere. This *second polar body* is good for discarding just like the first polar body. The other twenty-three chromosomes go to the ritzy side of town into a nice fat cell, the *definitive oocyte*. This cell, which contains most of the cytoplasm, also contains the intact nucleus of the captured sperm (see fig. 4.15).

The nucleus of the definitive oocyte and the nucleus of the captured sperm will form the basis for the baby as a result of their fusion. At this point these nuclei haven't joined together; they exist like two small golf balls inside a single egg. But the result is monumental. This is the first time in the entire process that two nuclei of opposite genders face each other as separate grocery bags of twenty-three chromosomes each. All biochemical attention is focused on them for the next series of processes.

Growth-Oriented Mergers

These nuclei react a lot like many people when they are center stage. They immediately swell up. Sensing this change, diligent scientists are aware that other names must be offered: the female swollen nucleus is a *female pronucleus,* and the male swollen nucleus is a *male pronucleus*. Swelling has started to occur because a real live genetic event is happening inside these pronuclei. The genetic event is a familiar one: these nuclei are duplicating their chromosomal material again. Now why on earth would they do that? They spent an inordinate amount of time re-

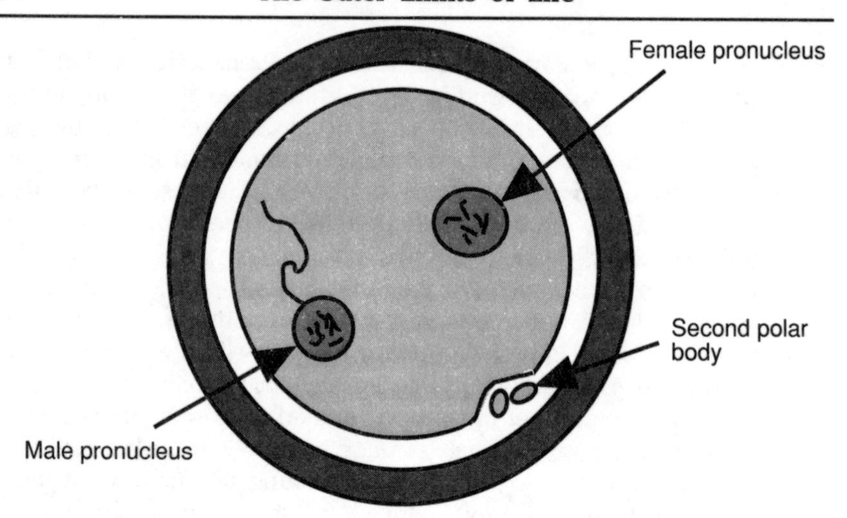

Fig. 4.15. **When the egg is penetrated by the sperm, it immediately undergoes the second meiotic division. This division results in the formation of the second polar body. The sperm has begun to degenerate even as its nucleus swells. In twenty-four hours the male and female pronuclei will have fused, and an embryo will have been created.**

ducing themselves to twenty-three chromosomes. They spent an even longer period of time waiting to meet. Why, still separate in their little golf ball enclaves, would they make forty-six chromosomes one more time?

They have to get ready to replicate as soon as possible. The goal of a baby is to make billions of cells from a doublet in nine months. By making double the amount of genetic information, the egg and the sperm will have enough genetic information to replicate as soon as they meet. The fertilized egg hits the ground running or, rather, hits the ground growing. It is a spectacular achievement. If an average six-foot human grew like an embryo does in its first month, he or she would be five times the size of Mount Everest.

This is how it happens. While the chromosomes are replicating, the golf ball-like nuclei are allowed to size up each other safe in their round sanctuaries for as long as twenty-four hours. Then, intricate little spindle fibers form like frost over each separate golf ball. When the chromosomal duplication is complete, the outer shells of the golf balls melt away. The chromosomes from each gender are thus exposed, like shy children at a junior-high dance, standing in a line facing each other. Chromosome one of the female soon pairs with chromosome one of the male. Chromosome two of the female pairs with chromosome two of the male. This pairing keeps happening until all pairs are lined up. This is the first joining of two people to produce a third unique entity.

Since double the amount of information exists (each cell contributing forty-six), ninety-two chromosomes are sitting there blinking at each other. But they are chromosomes derived from two separate persons. Soon that familiar earthquake happens, and this new mixture of forty-six chromosomes are trundled off to one side of the fissure and forty-six to the other side. Two cells are created from one in an act of pure mitosis. Thus, the first and foremost passion of the preborn cell is reproducing its own kind. No wonder we all have such an avid interest in sex. It has been our preoccupation since the day we were conceived.

Missed Conceptions

Given this knowledge, how do we answer the question regarding what is human life and what is human tissue? It would be nice if we could take these cellular miracles, put them into neat, tidily defined biochemical boxes, and sit back to wonder at the complexity of it all. The problem is that development, as we have seen, is escorted into our lives in progressive stages. We get our reproductive cells in the first month of our lives. But we do not get molecular permission to use them for many years. Even then, we can use only one cell at a time. When an egg and a sperm cell eventually meet, they do not form an instant nine-month-old baby. They grow processively, starting with two cells and letting time turn that number into 2 billion. Even the act of fertilization is not a singular one-reaction-and-it's-over event. A progressive series of biochemical reactions eventually leads to fusion of two nuclei that have never seen each other before.

Because this biological pageantry is incrementally developed, we have an excruciatingly difficult task ahead of us. We are forced to confront the constituency of life, and we must distinguish it from the constituency of human. We have seen that it is very difficult to talk about any biological process in monolithic terms. We had to be prepared for the biology to show us that life is a continuum, which is one way to cop out on uncomfortable biological facts.

A House Divided Against Itself, Part I

When we begin to tease away the genes and chromosomes of humans in their beginning stages, the issues get much more complicated. Fertilized human eggs don't look like us, and they don't act like us. We couldn't see their genes if it weren't for sophisticated laboratory equipment, available to us only in the last twenty years. Because of this abstraction, it is tempting to dilute the idea of protectable humanity into a continuum of tissue and ignore any consequences of trivializing who we are. If that, indeed, is possible.

This abstraction has caused many people to believe that a fertilized egg is only a potential human being, not an actual one. It is human tissue and not human life. In this view, human life is defined by its ability to perform certain human functions. If those functions are not resident, neither is the humanity. The fertilized egg cannot run because it has yet to develop legs. It cannot manipulate its terrestrial environment because it has not yet made an opposable thumb. It cannot think until it grows a brain. The potential biology of a fertilized egg is not the actual biology of a single two-celled entity.

This idea of unequivocal realities has an extension. Many fertilized eggs will never grow up to be anything. Different studies point out that of one hundred human conceptions, between 15 percent and 50 percent will spontaneously abort. There is no guarantee that any one fertilized egg will grow up to be much more than a normal menstrual period. For many persons, these arguments cast a shadow on the judgment that a fertilized egg is an absolute human anything. Instead, it is a potential human being, just like sperm and egg cells are potential human beings. In this view, the realization of a biological goal is not the same as its capability; that is, a marathon has not been completed because some of the participants had been judged capable of running it.

Many persons also subscribe to the view that a fertilized egg possesses nothing that we would call uniquely human. The biochemistry that powers the first weeks of human life is the same kind of biochemistry that powers the first weeks of chimpanzee life. Both are disturbingly similar to the biochemistry that powers adult yeast life. Because the pattern of biological development is not necessarily unique—at least initially—the separation between animals and human beings in this view is a very narrow one. Moreover, many animals as they leave the womb share lots of common characteristics with humans. They have collections of genes that make them breathe, force them to react to painful stimuli, cause them to get hungry, and so on.

Humanity in this view is bestowed upon a certain collection of genes only when those genes confer a property that separates them from all other species. The property that has been put forth primarily hinges on our ability to think in certain abstract ways. When an organism possesses sufficient complexity to be able to display those traits, an organism is said to be uniquely human. If it can't, it is said not to be uniquely anything.

Finally, there are those who believe that the potential for creating human life between eggs, sperm, and their fertilized counterparts is shared equally, with none possessing more humanity than another. Conception is thus no more important than free-agent sperm and unperturbed eggs. For example, it is true that all a fertilized egg requires to develop into a baby is a warm uterus and nine months. To perform an identical service, a sperm needs those conditions plus an available egg. An unfertilized egg needs those same conditions and an obliging sperm. The real reason that a baby can be created is not simply because of fusion but because of the composition of the individual cells that allow fusion to occur. The developmental program initiated at their union is only a product of that composition.

In this view, the potential for a key to unlock the door is not lessened because it has yet to be inserted. Humanity is defined not as the opening of the door but as the possession of the keys. The distance between eggs and sperm is considered to be an arbitrary distinction; conception should have no basis in determining what is human life and what is human tissue. So either we treat them all—eggs, sperm and fertilized eggs—with the same sacred respect, or we ignore their input equally. The extension of the view is that since we do not regard discarded sperm or eggs as lost humanity, neither should we regard their joining as anything more special.

A House Divided Against Itself, Part II

These ideas are fighting words to people who declare that life begins at conception. To many, life defined at any other point is a slippery slope toward the same philosophies that generated the killing fields of Cambodia and the concentration camps of World War II. At stake in this issue is nothing less than the personality of the bearer of the belief. Why? Because a common assertion is that where we think we begin tells us something about who we think we are. In this decade, geneticists are examining who we are at a level more intimate than anyone thought imaginable ten years ago. This issue has become an incredibly potent arbiter of contemporary human fate. It is not only adjudicating the futures of four thousand pregnancies a day; it is also determining the arrest record of some of our most law-abiding citizens.

And most important, deciding the political futures of our most prominent politicians.

Those who believe human life begins at conception would probably not be willing to eat dinner with those who believe that life begins at functionality. For example, if life is defined solely by recognizable human activities, humanity exists only when those characteristics are present and does not exist when they are absent. A human fetus cannot run like an adult. But neither can a five-month-old baby or a fifty-year-old paraplegic. A normal human is incapable of running when he or she is under anesthetic or is asleep.

Is humanity like some kind of Frisbee tossed between people depending on what they are doing? Although running may not be defined as unique to humans, life-at-conception advocates would argue that the logic is harsh even for some of our most unique unlearned traits. A one-celled human fetus does not have opposable thumbs. But neither do Vietnam veterans whose hands have been blown off by misfired grenades. Mentally retarded children cannot think with any kind of rationality or abstraction. Neither can some of our politicians.

Defining humanity solely on certain functional grounds means not distinguishing between these populations. And for those who believe human life begins at conception, a fertilized egg is one of our most defenseless citizens. Judging humanity based solely upon functionality is not only in error in this view; it is nauseatingly repulsive.

Such a strong opinion is not sequestered to functional equivalence arguments. In the conception view, trying to ignore the worth applied to fertilized eggs by examining spontaneous abortion statistics is just as enraging. The problem with the argument is that we do not excuse killing a group of people because we know that a certain percentage of them will die. We do not kill infants before the age of two simply because we know in advance a given number of them will succumb to sudden infant death syndrome. We do not kill fifty-year-old women simply because more of them than of thirty-year-olds will die of breast cancer. Attempting to devalue life because some humans will die sooner than others is absurd to those who hold life to be at conception. In their view, it is a lot like making it legal to wreck every car you see in a parking lot because you know that someday one of them will run out of gas.

Those who believe in life at conception strongly disagree with those who believe that the embryo is not much different from an animal. They refer to the fact that most gene sequences of humans are peculiar to humans. What these genes share in common with the rest of the animal kingdom is not the absolute identity of their nucleotide order but their functions. Since all terrestrial animals have to confront the same atmosphere, humidity, and relative temperatures, this is probably not surprising. Nonetheless, the genes that make the brain of chimpanzees do not have the same sequences as those that make the brain of humans. Neither do the genes that govern our body plan, skeleton, and hormone responses. The immune system genes embedded in a fertilized human egg are so specific that they will eventually be able to distinguish the developing human from any other person on the face of the planet, even a close brother or sister.

To add to this complexity, these distinct sequences are juggled into a combination at fertilization that is uniquely human at a whole new level. This uniqueness is primarily why we are spending $20 billion on the Human Genome Project and not the Avian Double-Breasted Sapsucker Genome Project. So much can be learned by understanding the uniqueness of human genes, we are in the midst of sequencing our forty-six chromosomes. For those who hold the view that life begins at conception, there is more than enough molecular uniqueness to distinguish the human organism from any other organism.

Finally, many take issue with the idea that a fertilized egg is no different in its humanity from a sperm or an egg. In this view, the initiation of the developmental program as a result of fertilization is the unique human-producing event. The players are necessary in that the union could not occur without them. But it is held that neither egg nor sperm has the capacity to bring about the formation of the preembryo by itself. The combination, not merely the presence, of egg and sperm is the necessary and sufficient condition to separate human life from human tissue. In this view, an automobile is observed moving down the freeway not because someone possessed a car key but because someone inserted it into the ignition.

Corrective Bargaining

The sad thing about the debate over the origins of human life is that it will only become increasingly hostile as biochemical discoveries reveal more about us. We are already finding that many arguments formulated on this question have relied on incomplete information. Many of the controversies swirl around the value of the unique entity created at conception, the ability of a sperm to fertilize an egg. As we shall see in the next chapter, the very fact that fertilization is required to create another human is now being called into question. And neutral science, with discoveries like that, will keep nonneutral humans stirred up for a long time to come.

We have always run the risk of arming our value judgments only with scientific facts. The danger is that we end up substituting data collection for ethics, as if the knowledge in our heads could imitate the values in our hearts. This is why people have double reactions to science, either wanting to stop every experiment for fear of the next Frankenstein's monster or wanting to allow everything for fear of the next Joseph McCarthy. That's a modern error.

Simply because science can explain many amazing things does not mean it can explain all amazing things. We cannot keep allowing our intellectual abilities to seep into our emotions and characterize our moral judgments for us. We will eventually empower ourselves to make decisions about issues for which little ethical thought has been brought to bear. This will allow us to treat one another like we treat test tubes, conferring on ourselves the ability to tell one another, and to tell God, what we will and will not do with the infinite miracle of procreation. We will become as absurd as a little second-grade boy I once knew who informed his teacher that though he didn't need her help for that academic school year, somehow she might need his.

Chapter 5

The Breed of Life

Introduction

In 1984, the Warnock Commission was created in Great Britain to study the feasibility of performing experiments on human embryos. It recommended that carefully regulated experiments be permitted on human embryos up to fourteen days postconception. The recommendations of this report formed the basis of the Human Fertilization and Embryology bill which, if passed, would formally codify the practice into law. The bill was mostly noted for the controversy it engendered. In February 1990 the House of Lords passed the legislation, and in April 1990 the House of Commons passed it. The advocates of the bill believed such research represents the highest callings available to humankind. The detractors believed such research incites some of the darkest deeds we are capable of doing. You can guess that it was not a friendly controversy.

The controversy focuses on the biology of early events in embryonic development. As we will see in future chapters, such development consists of a very few genetic themes played out in almost infinitesimally complex variations. These developmental themes are laid down early in the life of an embryo; the variations occur as it grows and develops. That is why genetic malfunctions early in development usually have such disastrous consequences. If the resultant baby survives the malfunctions, it is usually horribly deformed. The biochemical reactions in the first two weeks thus represent some of the most fundamental events in the human experience.

A Touch of Glass

Performing experiments on human embryos during those initial stages of development is an extremely tempting line of research. Isolating the genes that control early events can lead to a greater understanding of virtually every known human birth defect. Future generations of experiments can focus on fixing the molecular switches or following a particular series of switches until one of them fails. Procedures that will allow genetic intervention on these defects represent real ways of conquering them. Without these experiments, we are forced to make guesses about human life based on the use of animal tissues and certain kinds of human cancers. Standing idly by while babies are born with crippling diseases and lethal grotesque malformations is an awful feeling for a researcher.

As noble as all that might sound, these manipulations remain extremely controversial. The unique peculiarity of genetic engineering is that one moral strength can be pitted against another based solely on technical capability. For example, most people would not argue that critical biological information could be obtained by performing such experiments. Nonetheless, acknowledging this biological good says nothing about the content of the protectable humanity within the experimental subjects.

A familiar battle is thus rejoined. If a fertilized egg is a human being, allowing experimentation on this entity has the same moral equivalence as working on an adult. In this view, it does not matter whether the human weighs a fraction of an ounce or 120 pounds. Performing experiments to induce malformations, weird recombinations, or whatever is necessary is tantamount to torture in its most hideously permanent form. Destroying a human life after the experiment is over is a murder so dehumanizing as to be worthy of our worst crimes. We have already made a judgment, from Nuremberg to Mylai, that trivializing human life is morally impermissible. For those who believe human life begins at conception, the moral cost for these research benefits is so high that paying it would bankrupt us.

Tissues and Answers

The controversy continues. The debate over the bill in England represents only the latest skirmish in an increasingly widespread war. Sweden already allows research on fertilized eggs up to fourteen days postconception. West Germany has passed a law banning such research.

As usual, these controversies collapse into a single question: What is human life, and what is human tissue? The Warnock Commission's recommendation aims this question right at conception. In the last chapter, we examined some of the biology of that process. Now we must use this knowledge to address the finding that the process of conception is not the only way to create functional embryos. It may not even be the only way to create functional human embryos. When one examines the biochemistry of the process, one is impressed with the incredible flexibility of most aspects of conception. If one is going to make moral decisions about particular experiments to perform, one will have to take such pliability into account.

We will first examine the animal world, specifically investigating the reproductive strategies of organisms that reproduce with eggs and sperm. Next, we will examine the flexible roles of gender in the production of those eggs and sperm, especially in organisms that only reproduce sexually. Finally, we will examine whether the joining of eggs and sperm is mandatory in the formation of viable embryos, including human embryos. Any definition of human life must consider the fact that the processes of its formation are not monolithic. It must also consider that such flexibility is observed, like a reflection, in the glass cage that houses the animals of God's creative zoo.

Gross Encounters

Let's first investigate the manner in which certain creatures get their eggs and sperm in close enough proximity to be useful. The animal kingdom has almost as many methods of reproductive strategies as it has organisms. This is even true for organisms that reproduce sexually, utilizing egg and sperm. Some animals have extremely odd ways of making sure the egg of the female reaches the sperm of the male. Some animals start out as one gender and then later in life turn into the other gender. Some animals are both male and female at the same time. Some animals do not need an opposite gender to reproduce. This topsy-turvy world illustrates that definitions of genders—and the embryos they create—are about as set in stone as yesterday's stock market report.

Our first example of odd reproductive strategies involves the deep-sea-dwelling angler fish. This funny-looking, frying pan-sized creature is a member of order Lophiiformes. I describe here the female of the species. The male is a boring little fish, a couple of inches long at full growth. He spends much of his time trying to find the female in the pitch-black depths. When he does find her, the male will attach himself to the female's body by jaw and suction and hang on for dear life. In this position, the eggs and sperm share the same space of ocean, and the organism can reproduce.

This freeloading affair, though convenient at the beginning, ends up exacting a terrible price from the male. While he is hanging on, the female secretes special enzymes that permanently fuse his mouth to the surface of her body. Blood vessels from the host female invade the hapless hitcher's body and destroy his brain; the male dies in this embarrassing position. At least one fisherman has told me that you can roughly assess the age of a female anglerfish from the number of former lovers attached to her abdomen. These fish start out as two separate sexes but end up as a single animal with both.

Another odd mating ritual comes not from a fish but from an animal fish like to eat: a worm. The water-dwelling *Bonellia* worm (also called the star worm) is a tiny organism that looks more like a floating pickle than a biological organism. The female has a dumpy warty body with an extremely long nose. The *Bonellia* worm doesn't start out looking that way, however. It starts out as a tiny nondescript and genderless larva.

The larva is small enough to fit in the female worm's long nose. If the larva encounters a grown female, it will unceremoniously enter her nose. The female will shower the intruder with certain hormones, and this bath will turn the larva into a male that will never get any bigger than a tiny dot.

The male takes revenge because of this transformation by sliding down the gut of the female worm. It then travels inside her until it encounters her reproductive organs, where it sets up a permanent home. The first egg that slides down a tube called the *oviduct* is fertilized by the little male. Thus, the female becomes permanently pregnant by creating a male inside her nose and then swallowing it. Kind of like the opposite of birth control pills.

There are many curious strategies for getting germ cells together. No matter how odd they are, however, fertilization can occur in these animals because of fixed absolute genders. But are there examples in the animal kingdom where gender is not so set in stone? It would be a foreign idea to most of us that gender might not be a permanent addition to the life-styles of all organisms. Once a male, always a male, and once a female, always a female. Right?

Well, no. Biology once again escapes the confident claws of the monolith. Numerous examples of organisms change genders in their lifetime; some change genders as easily as humans change clothes. To understand how this occurs, we will examine the life cycle of a certain snail.

Gender Benders

Slipper snails consist only of a single gender when they are born; all are male. One might ask how any of the species survive if they are a bunch of young guys floating in the ocean. The answer, of course, is that they do not remain guys permanently. Their masculinity is relatively short-lived.

When the slipper snail attaches to the back of something, anything, it immediately begins to turn into a female. If a young male is fortunate enough to attach to the back of a female snail first, he will mate with her before he completes his transformation. But eventually he will turn into a female. If another male attaches to the back of this newly transformed female, he will mate with her before he is transformed into a female.

Over time, the scene becomes ridiculous. A sexual tower of snails fourteen organisms high is not uncommon. All the bottom organisms are females. All the top organisms are males. All the organisms in between are males turning into females. Gender flexibility is easy to study in these organisms because all the intermediate types are so readily available.

Not all gender-switching organisms start out as biological males and then go through a gender crisis, however. Some gender-flexing organisms start out as females. Numerous organisms were quite mysterious to researchers because only females could be found. It was later discovered that the males in one of these organisms appeared only when a severe drought or other thermal environmental stress occurred.

Did they come like white knights to rescue the reproductive future of the species? Quite the contrary. It turns out that some of the females could transform into males during an alteration in the environmental temperature. If it got too hot or too dry, some females changed into fully functional males, which meant that sexual reproduction could take place. It also turns out that the males were smaller and ate less. The species survived better if some became masculine because there was less stress on an already troubled environment. It's kind of like going on a diet by converting into an organism with no appetite. In this case, males were used as a defensive reaction to a change in living conditions. When the environment settled down, the normal population of females resumed.

This sexual versatility has created interesting partnerships in the animal kingdom. Some organisms like the gender models so much that they do not switch roles. Instead, they have been genetically programmed to keep both sexes within one body. Such creatures are called *hermaphrodites*. Many such organisms have retained the ability to self-fertilize, thus solving the age-old problem of trying to find a mate. Representatives of hermaphrodites range from worms to fish. Humans are capable of a kind of hermaphroditism, although these people are usually sterile.

Mixed Doubles

One example of productive hermaphroditism is the pork tapeworm, a common parasite in the intestinal tract of pigs. It looks like a series of small flat pillows that have been sewn together into one giant chorus line. The first pillow, called a *proglottid,* consists of the head. Each of the other pillows contains exactly two organs, a penis and a vagina. Each penis has the ability to fertilize its own vagina. If the tapeworm is folded up onto itself, the penis of one proglottid can insert into the vagina of another. If two worms are side by side, multiple fertilizations can occur between the two organisms. These self-fertilizing tissues eventually form eggs that can be released into the abdomen of the host pig. The tapeworm is the ultimate mating machine: one giant organism impregnating itself literally hundreds of times a day and shedding its eggs into its environment.

This reproductive strategy has an advantage. These organisms exist deep in the interior of the pigs, an environment not exactly conducive to matchmaking. Being fortunate enough to get a double infection with one of each gender in a single pig would be a rare event. Far better for the survival of the organism if it could carry all the reproductive apparatus it needs to form its unions internally. And that is exactly what the pork tapeworm does.

Another example of hermaphroditism is a perch that lays its eggs in the bottom of a lake and then hovers over them, spraying the brood with its own sperm. It performs external self-fertilization. Other organisms look like males all their lives. But buried deep inside their testicles is a little ovary that can produce an egg. At the right time, this egg is released and automatically fertilized. The male either becomes pregnant or releases the fertilized egg into the environment.

One of the most extraordinary examples of semiautomatic impregnation occurs in leeches. In an organism called *Gyrodactylus,* an adult will carry a juvenile around in its uterus. No big deal. Lots of pregnant organisms do. But this internal juvenile also carries another juvenile around in *its* uterus. The normal arrangement for the life of these organisms is a box within a box within a box arrangement. A single act of sexual intercourse is enough to impregnate all three organisms. The organism ends up storing the sperm from such copulatory acts in its body. Future generations can become fertilized without the necessity of external encounters. This masculine transference is really not hermaphroditism; delayed fertilization takes place because of the external deposit of sperm.

What can be learned from these organisms in terms of our view of the beginnings of life? What does a discussion of these processes have to do with our ideas of conception? Specifically human conception? The answer is nothing. Yet. These examples merely illustrate the biological fluidity of reproductive strategies and genders. Because biological variability exists with the reproductive participants, and thus at the very start of the process, an interesting shadow is cast on the rest of the routine. We see that it is wrong to think of sexual identities in terms of fixed, inviolate biological states for all organisms. One is tempted to ask if any other processes in reproduction are just as fluid.

For example, is it wrong to think of the process of fertilization just as rigidly? In our discussion so far, we have not encountered any reason to think so. Or to think not. Are there natural ways of creating embryos that do not involve the joining of sperm and egg? Can these pathways be induced in the laboratory? Would this confer on us the ability to create embryos without the use of fertilized eggs? The answers to these and other questions are described next. To start, we must understand a process scientists have termed *parthenogenesis*. Probably so named after the discoverer visited Athens.

Parthenogenesis: the Ultimate Me Generation

What is parthenogenesis? This process lays to rest forever the idea that all biological organisms that reproduce sexually need both eggs and sperm to make babies. Simply stated, parthenogenesis is the ability to create an embryo without fertilizing the egg. You did not read that wrong. Certain mechanisms allow embryos, even full-fledged organisms, to be created without fertilizing anything. The sperm under these conditions does not genetically contribute anything; it doesn't even need to be around.

These mechanisms occur naturally and can be found in organisms as simple as aphids and as complex as birds. Parthenogenesis can be induced in the laboratory, and functional embryos can be created from simple eggs. Some researchers believe that parthenogenesis can be induced in humans; the only reason we haven't done it in the laboratory is that we haven't tried.

If parthenogenesis is real, we will have to change the way we think about conception. Or the way we view what is experimental tissue and what is not. And that brings us around to an extremely familiar issue. We have to rethink what we believe about the beginnings of human life.

Let's talk a little bit about the biology of parthenogenesis. This process can occur in a surprisingly large number of ways. You recall meiosis, complete with its two divisions. Do you remember that when an egg completes its first meiotic division, it usually produces one normal-looking egg and one egg full of cellular junk? You might recall that junked-up egg is the first polar body. Parthenogenesis can take place when that first polar body is formed and then, for some reason, begins to think it is a sperm cell. This renegade polar body will then collapse back into the normal egg cell. The unexpected fusion results in the egg's believing it has been fertilized by a sperm. It will undergo the second meiotic division spontaneously, only with double the amount of chromosomes because of the collision. And then it will make an embryo on its own.

Parthenogenesis can also occur without fusion or help from any obvious source. A normal oogonium containing a normal amount of chromosomal pairs might look, from all outward appearances, perfectly average. It should go through meiosis, at least through some of its initial stages, and wait patiently for its wiggling little masculine suitor. For incompletely explained reasons, this egg loses patience and bypasses meiosis. Instead, it goes through several rounds of replication (it even forms a polar body in the process), and when it's over, an embryo has started to form. No intercourse, no activation, no sperm. Simply an embryo derived from an egg.

The animals that undergo parthenogenesis in the wild consist of both invertebrates, which are wimpy creatures without a backbone, and vertebrates, which are wimpy creatures with a backbone. Humans are vertebrates, for example. Because of the large number of participants, some researchers wonder at the necessity for any male contribution at all. A thought no doubt shared by many women throughout history. This biology, despite its social consequences, points out that the developmental program necessary to create a complete creature is not exclusively resident at conception. A life can begin by the appearance of an egg.

Parthenogenetic Creatures Have No Backbone

So which invertebrate organisms undergo parthenogenesis? Lots of them do. Certain grasshoppers do not have a known male counterpart. They undergo a type of meiosis that doubles the number of chromosomes prior to the second meiotic division. And you guessed it, they produce only daughters. Certain types of fruit flies undergo parthenogenesis. So do some species of brine shrimp. Some butterflies do it, and certain kinds of worms do it. There is an aphid that will reproduce by normal sexuality in one generation and in the very next generation will be parthenogenetic. In one generation both males and females will be produced, and in the next generation only females will be produced.

Perhaps the most stunning example of parthenogenesis in the invertebrates is that of the common honeybee, *Apis mellifera*. When a normal queen matures, she escapes the hive and takes one of several nuptial flights. Several male drones follow her into the air and eventually mate with her. The sperm that she receives during this high-flying union will be stored in a special pouch, the *spermatheca*. This supply will last her the rest of her life.

When she comes back to build her hive, she can lay one of two types of eggs, fertilized or unfertilized, apparently at will. The eggs that are fertilized will dip into that supply of sperm kept in her pouch before she lays them. They will become either the workers or other queens. Eggs that are to remain unfertilized will not dip into that pouch. Those unfertilized eggs will become bees anyway; they will develop into drones as a result of parthenogenesis. Fully developed male honeybees have no genetic father, only a royal mother.

Parthenogenesis represents an interesting way to confine genetic information to a single source. The process can be used to create either gender and, in the case of the honeybee, both genders. If parthenogenesis was confined to invertebrates, however, it might never enter into a discussion about human life. But the process is also seen in complex vertebrate creatures. An animal whose fertilization requirements can be described as intermediate between true sexuality and parthenogenesis has been observed. Curiously, it is a tropical fish.

Fish Switches

This tropical fish, the Amazon molly, has never been found naturally or otherwise in the Amazon and has been found naturally or otherwise only in Texas and Mexico. All of its offspring are female, which demonstrates that the scientists who named this fish were more familiar with Greek mythology than with geography. The lack of male presence was an immediate clue that parthenogenesis was occurring or the male was some kind of unexpected creature.

These fish undergo a sort of parthenogenesis. Sperm are not required for genetic input; the female does well enough on her own. However, she needs sperm to jump start the embryo-making process. The female uses the sperm of other species to make her eggs hatch. They do not donate genetic information; they do not penetrate so that fusion can take place. And any old sperm from any old species of molly will do. The developmental program is fully available in these eggs. The eggs need the sperm for final growth only like a car might need jumper cables to travel.

The molly is an example of an intermediate-bridging mechanism between normal fertilization and true parthenogenesis. The real stuff, the complete absence of any male input, has been found in a wide variety of complex vertebrates: amphibians, reptiles, and chickens. Sometimes this mode of reproduction leads to competent adulthood. Sometimes the organisms go only as far as eight-celled embryos. Regardless of the stage of growth, the familiar theme of embryonic development without fertilization is encountered.

Parthenogenesis is also seen in chickens and turkeys. The first description of an unfertilized egg undergoing the initial cleavage stages of a real embryo was reported over a century ago. It has since been demonstrated that a wide variety of chickens undergo parthenogenesis. Some go through it rather frequently; in one species where, one egg in twenty develops into an embryo without sperm. Among certain species of turkeys, that number goes up to almost one in five.

Not all of them grow up to be adults, however. Those that do can display unusual genetics. Some grow to be adults with only one-half the genetic blueprint necessary in normal counterparts. Some grow up possessing triple the amount of their genetic material. These extra chromosomes do not appear to get in the way of most aspects of normal life. The implications of this find are somewhat stunning: the normal amount of chromosomes is not a prerequisite for growth to adulthood—not to mention that neither is the input of sperm.

The Mice Is Right

The startling thing about parthenogenesis, besides the fact that it occurs at all, is that it occurs in so many different species. The more complex a species becomes, the less likely the process is to take place spontaneously. Even if it happens spontaneously, a survivable adult is less likely to be created. Nonetheless, parthenogenesis has been demonstrated in higher organisms such as mammals. It has occurred in good old laboratory mice and their larger cousins, the laboratory guinea pigs. A special strain of mouse has been developed to study the process. This female routinely produces eggs that undergo parthenogenesis about 30 percent of the time. None has developed to full adulthood.

Parthenogenetic processes can already be induced in eggs that would normally require a sperm for development. This work has primarily been done in mammals. The process can be induced a number of ways. For example, treating normal mouse eggs with a fair amount of heat can induce more than half of them to start creating embryos. Treating normal rat eggs with sudden cold can transform almost 100 percent of the unfertilized cohort into partially developed baby rats. You can dip an egg into a dilute solution of alcohol, and the result is an egg whose developmental program has been initiated.

The best results are achieved with the application of strategically placed electric shocks, which has to be done surgically. The female mouse is anesthetized and opened to expose her reproductive organs. Tiny electrodes are placed on either side of her ovaries, and an extremely weak electric shock is applied. This shock is administered gently to prevent discomfort and to avoid burning the ovaries with a large pulse. Once the process has been completed, the ovaries are removed and the eggs examined. Almost three-fourths of the eggs have started on the road toward becoming baby mice.

Whether it be alcohol or electrical shocks, parthenogenesis can be artificially induced in vertebrate animals. This induction has yielded information about the developmental program and chromosomal number. Like the chickens, some mouse embryos have only half the normal genetic information. Other embryos have double the amount of information. A few have four times the amount of information. The developmental program is nonetheless incited in these apparent mutations. This program appears to have flexibility in its requirements for specific amounts of genetic information.

Human Eggs

If we are moving up the ladder of increasingly complex organisms, we have to ask a very painful question: Does parthenogenesis occur naturally in humans? Would we be ready for the implications if the answer was yes? The answer to that question, according to a number of reports, does turn out to be yes.

Parthenogenesis can occur naturally in human females. A four-cell-stage human embryo was described some time ago. Two facts demonstrated that the embryo was not the product of an unaccounted act of normal fertilization. First, the beginning growth stages of the human embryo were not observed in the Fallopian tubes, where fertilization normally takes place. It was found deep in the ovary. Second, the embryo was isolated from a seven-year-old girl. She was not pregnant, at least not in the traditional sense. She had not undergone puberty. This egg, already primed, did what all parthenogenetic cells do. It short-circuited its normal developmental clock and, in this case, skipped over the watchful eye of ovarian inhibitory substance. For reasons totally unexplained, that was enough to incite the beginnings of development.

This phenomenon has now been reported in other human females. Though we are more complex organisms than most animals, we share many things in common with them. One of those characteristics is the ability to begin making preembryos without first conceiving them.

From a genetic point of view, the next most painful question that must be asked is this: Can parthenogenesis be induced in human eggs? The answer to that question, in the opinion of many leading researchers in the field, is also yes. Probably. That human eggs can spontaneously undergo cleavage demonstrates that the initial developmental program is preset without needing masculine input. At least in part.

This disturbing fact means two things. First, all the information necessary to make a baby is available genetically in the egg alone (more on that subject later). Second, this baby-making information has been lined up in some kind of order, nearly ready for work. It is missing a developmental start signal, normally provided by the sperm. Somewhere in these preembryo-carrying human females, that signal was provided. And that's the clincher. Finding that signal will mean that human embryos progressing to some stage can be created without fertilization. At will. That is why many leading researchers say that deliberate parthenogenesis can be a reality in human genetic research.

The Way We Blur

As more and more countries give permission to perform experiments on human embryos, mechanisms like human parthenogenetic development will be better understood. So will a lot of other developmental mechanisms. The insights gleaned from such research will be at the frontier of modern biological thought, complete with the attendant border skirmishes with ethics. This is true only because it is impossible to carry out legally, not because it is impossible to carry out scientifically.

All of this input brings to mind a disturbing aspect about the harsh light of molecular biology. Whether we are speaking of reproductive strategies, gender determinations, or mechanisms of fertilization, processes that we have always assumed to be rock solid often turn out instead to be gravel. Just when we think that gender is a fixed state, science shows us organisms that have both genders or neither gender or that can freely change from one gender into the other. Just when we think that life begins at conception, science shows us a way to start life without conception. Such data can chip away at the time-honored cages that capture our ideas about biochemical processes and ourselves. In the long run the accuracy of any information is preferable to the errors of comfortableness. Being incarcerated by actual facts is much better than being bailed out by scientific mythology.

Assuming that someday it will be possible, we might be tempted to say that parthenogenesis is just another way to make humans. The definition of conception would be stretched to include this alternate way of inducing the developmental program. Recent biological research suggests, however, that preembryos can be created in tissues that don't look very much like eggs at all. Or sperm. Instead, they are created from pieces of embryos that are able to make organisms, like the pieces of broom handles that make entire brooms in *The Sorceror's Apprentice*. Thus, the process of conception turns out to be much more flexible than anybody thought. And the recommendations of the Warnock Commission guarantee that this research will be unleashed on human embryos not more than two weeks old. The question then becomes, How do we define the humanity of such cells? The consideration of this topic is the subject of our next chapter.

Chapter 6

Womb with a View

Introduction

I've always been amazed how teenage groups tend to cluster, like snowflakes form around dust particles, with friends who own cars. My teenage years were no exception. The gang I hung around with had Dave as the honorary car owner. He drove a really hot Chevy Nova, an automobile he had inexplicably named Passion Semester. We would travel everywhere in Passion Semester, often with nothing more particular to do than watch the same street corner go by a hundred times in one evening.

I remember one summer's night we were busy collecting the gang together to review the status of our street corner. Cramming about five of us in the car, we sped off to get one more of our troupe. George lived in a new part of town, an area filled with uncompleted construction projects and neighbors who let their giant dogs run free in the neighborhood. When we parked near George's new house, Dave had to maneuver Passion Semester around a rather large collection of discarded boards to get to the driveway.

As is true of teenage tribal behavior, we all began to pile out of the car to get George. Dave was in the lead (his natural position), car keys already stuffed in his pocket. As he neared the house, a giant German shepherd nearly the size of Passion Semester suddenly howled from the far end of a half completed house. We, who were still in various stages of getting out of the real automobile, heard this cry and watched in horror as the dog began leaping over the piles to get at us. With typical middle-class American courage, we fell over each other climbing back into the safety of Passion Semester's womb. We then locked the doors.

All of us, that is, except Dave.

Dave was too far away. His keys were in his pocket. His car doors were locked. His buddies were the consistency of vanilla pudding. And a Heat-Seeking Stealth Killer Dog was barreling at him at nearly the speed of sound. What could he do? What was there to do? He quickly surveyed the area, found the convenient pile of boards, and grabbed a two-by-four.

Then lifting the board high above his head, he yelled at the top of his lungs a quite biologically correct description of the hound's relationship to his mother. Then Dave ran, full speed, right at the dog.

And what did the dog do? The dog, stunned at this extremely foreign behavior of a white middle class male, stopped dead in his tracks. He appeared to look at this apparition in a state of shock. The dog then let out one more great big yelp, turned around 180°, put his tail between his legs, and took off in the other direction. There was Dave, brandishing his board like it was Excalibur, in hot pursuit of a six-ton hound-from-hell. As he went bounding over the construction site, we took courage and got out of the car. That was just enough time to hear poor Dave yell, as he disappeared over a small hill, "Hey, guys!! What do I do now?"

I will never forget Dave and that dog. In just seconds after he had vanished over the hill, we again saw Dave running for all his life toward his car. The dog was in hot pursuit, with Dave's board clenched tightly in his jaws. Dave had confronted the very thing that appeared to threaten him, and once it responded, the situation almost grew to be more than he could handle. It would not be the only time I was to watch someone nearly get over his head in a fearful situation.

I could almost smile at it then, mostly because it hadn't happened to me. When it was over and the dust dulled our memories, we would laugh "with" Dave (he made sure it was never "at") about that dog. But I would not always be smiling when I had to confront certain fearful situations myself. Later in academic life, I was to find that there is often very little laughter in the midst of uncomfortable and sometimes morally ambiguous situations. Especially when the hounds would come not from newly constructed houses but from newly constructed ideas. When I cloned my first gene (which was from a simple fungus), I realized the tremendous power of certain ideas. Especially when ideas could reduce living organisms to test tubes.

The reality of research is sometimes a mixture of almost comical misadventure and electrifying, every once in a while terrifying, discovery. We have already seen that fertilization need not occur to create functional embryos. Some organisms require only an egg. Are there any other ways to create functional embryos besides normal fertilization and parthenogenesis? This chapter will attempt to give a partial answer to that question. We will first turn to the animal world to understand how the marriage of embryology and molecular biology has begun to unlock the secrets of development. We will then explore ways such research has been extended to human biology. Finally, we will consider the implications of such research on the issue of sentience versus tissue. The implications in this particularly controversial set of ideas will place us in the middle of a molecular construction project. Whether we get out of the car may depend on the size of the dogs we encounter once we arrive.

Breeding Between the Lines

Genetic issues have not settled comfortably into the collective psyche of twentieth century history. Whether one talks about the eugenics movements after World War I or the genetic excuses for wholesale slaughter in World War II, the memory is painful. Even without the history, grappling with modern genetic experiments is often difficult because the results are so terribly personal. You cannot have two more volatile issues to talk about than sex and death. Genetics is vitally wrapped up in both processes.

Perhaps this sense of foreboding exists because we are learning more and more intimate details of those processes. The rate of accumulated knowledge has been accelerated by the wedding of two very powerful disciplines of biological science: classical embryology and molecular biology. Embryology has been dropping bizarre and fascinating hints about development since the turn of the century. Molecular biology has picked up the scent. It is translating the embryological data into solid genetic and chromosomal mechanisms. The isolation of those genes and chromosomes means that those mechanisms will become available for human manipulation at will.

The fear that this double-teaming of embryology and genetics evokes in people ranges from the thoughtful to the ridiculous. National tabloids, as absurd as some of their conclusions are, in many ways appear to reflect a legitimate national distrust. And an incredible misunderstanding. I have found articles that describe

- The cloning of Hitler
- The cloning of Michael Jackson
- The gender switching capabilities of Michael Jackson
- The cloning of Elvis (from artificial laboratory cultures whose cell source comes from a piece of belly button lint obtained by a fan after a concert)
- The cloning of Elvis's voice (given as a gift to another person from, of course, beyond the grave)

I have also run across more legitimate journalism that has posed lots of "what if" questions about future embryological research. Some journalists like to speculate on the ethics of giving juvenile athletes the bodies or athletic capabilities of famous sports stars. Some have speculated on technologies that would give humans certain desired traits, like the outer physical appearance of movie personalities. Other journalists focus on the unique benefits of regrowing Einstein's brain, presumably by placing it in a person who had none.

Multicultural Experiences

One does not need the creative imagination of tabloid editors to understand that great advances have been made in the fields of embryological science. The tabloids are correct(!) in reporting that embryos can be cultured artificially in the laboratory. That is, fertilized eggs can be taken out of animals and placed into round petri dishes. If the conditions are right, the dishes will support the growth of little embryos. The requirement for successful culture dictates that these cultured embryos develop normally when they are replaced into animals. Thus, the embryo is formed in the animal, grown for a while in a dish, and then put back into the animal. It is, of course, possible to fertilize the egg in the dish these days and grow the little embryo outside for a period of time. More and more animals, particularly agriculturally important ones, can be grown in that fashion. New systems are being developed to make the process of such culture available from any animal source.

One system is the *co-culture technique*. Embryos are grown artificially under very specialized conditions. Clear round plastic dishes, about the size of small pancakes, are generally used. In the laboratory, the dish is flooded with liquid. A special type of cell called a *feeder* is then seeded into the dish. Feeder cells have been gently removed from an animal's uterus or ovary. Once in the dish, these feeder cells settle to the bottom and begin growing, eventually making a living sheet on the surface of the plastic. It forms an artificial womb that can support the growth of embryos. Special liquid is added so that these feeder cells have a constant supply of dinner.

The embryo is then introduced to the dish. The feeder cells keep slurping up the nutrients provided and give back to the embryo chemicals that allow it to grow as in a natural womb. With such a technique, fertilized eggs from various animals can be grown to a fairly advanced embryonic stage. Eventually, however, these cells must be transferred to a sexually mature female if full development is to take place.

The ability to culture embryos outside the womb has allowed researchers to find out a number of hidden secrets about their development. For example, cow embryos can be grown to a certain stage and then cut in half. Each half of this embryo is then placed into separate surrogate mother cows.

Surprisingly, the process of splitting does not kill the little embryonic bovines. Instead, each half will grow back its missing cellular parts inside the surrogate cow; identical twins are derived from a single split egg. This demonstrates that the genetic information necessary to create a completed embryo is preserved even if it is cut in half.

Out on a Lamb

Researchers decided to see what minimal cellular unit would create a complete animal. As you recall, a normal fertilized egg begins as a single cell. Inside that single cell are fused nuclei, one part derived from the female and one part derived from the male. You also recall that the first thing this cell does is to divide into two cells. And then it divides into four cells and then into eight cells and then into sixteen cells and so on. Perhaps reflecting a deep-seated desire to be munitions experts, scientists have named these cells *blastomeres*. Each blastomere contains its own nucleus and is a single cell.

An experiment performed in the early 1980s demonstrated the genetic power of individual blastomeres. It was done in sheep. Two cellular players were involved. The first was an embryo exactly eight blastomeres big (often called the *eight-cell stage*). The second was an ordinary unfertilized egg. The unfertilized egg was cut in half, like the previously mentioned embryo, in a special growth medium. Its nucleus was extracted in this procedure, leaving the cell as a round remnant with no genetic material. Such an egg is *enucleated* (probably to demonstrate that the letter *e* means something to scientists that it means to virtually no one else). Then exactly one of the eight blastomeres was taken from the intact embryo. The other seven were thrown out. That nucleus-deprived cell was fused to the single blastomere with a very specialized virus. The quasi-single engineered cell was implanted into a specially prepared female sheep. To just about everyone's surprise, that cell grew into a normal lamb.

In this experiment, all the chromosomal information was derived from one cell of a developing embryo. The unfertilized egg, stripped of its genetic information, provided a molecular soil in which the lamb could develop. This result, in the context of similar data obtained with other animals, clearly demonstrated a frightening biological reality: conception was not the only way to grow a new animal. A single cell derived from an already developing embryo was equally satisfactory.

The Roll of the Mice

This result caused certain scientists to wonder if an intact cell was necessary to create a living animal. Was it possible to tear down an individual cell, take out its nucleus, place it in an enucleated egg, and get an intact embryo?

The researchers decided to conduct experiments with mice. As before, two types of cells were used. First, the garden-variety unfertilized egg had its nucleus removed. It would once again be the soil to support the growth of an organism. Because its nucleus was removed, this egg could contribute almost nothing genetically to anything that got inside it. The second type of cell was a one-celled fertilized egg instead of an eight-celled organism. A sperm had already joined to this egg and had dumped its nucleus inside it. You probably recall that this masculine bundle of genetic information sitting inside an egg is a pronucleus. In this experiment, the male pronucleus had not yet joined to its female counterpart. It was thus a genetically distinct and separate bag of chromosomes.

Using exquisitely sensitive surgical techniques, the researchers took that male pronucleus out of the fertilized egg before it could fuse to the female. Because it did not fuse, it had half an embryo's normal genetic information. With even more sensitive techniques, that pronucleus was placed into the enucleated egg. Thus, a nucleus, instead of an intact cell, was put into the enucleated recipient. The entirely human-engineered cell was then placed into the uterus of a specially prepared female mouse. Would this construct, one that never saw conception, nonetheless turn into a baby mouse?

The disturbing answer is yes. Repeatedly. In this experiment, all the chromosomal information was derived from a single nonfused nucleus. Not a fused nuclei, not a cell from an already developing embryo, just a bag of chromosomes. The unfertilized egg, stripped of genetic information of its own, once again provided a bed in which the baby mouse could grow. But a bed was all that it could provide. The same reproductive flexibility observed in the lambs was repeated in the mice; conception was not the only way to grow a new creature. A single nucleus derived from an undeveloped fertilized egg was enough to create an entire animal.

A Dish Out of Water

These experiments show an astonishing flexibility in the power of individual cells to create entire creatures. However, the tendency is to over-interpret the animal results as we seek to generalize the implications. Enormous technical difficulties remain about most of these experiments when considering the animals themselves. It is difficult to export this work to more complex organisms.

Nonetheless, can such research be applied to human beings? Are human germ cells as flexible as those of other animal species seem to be? If you burst apart the cells, could entire humans be grown from partial fragments of embryo? The answers to these questions await experimental evidence except in one case. In this case, an experiment has already been performed on humans that says the answer is probably yes. To understand how this is so, we must review human embryology and discuss a technique you may have read about extensively in the newspapers, the technique of human *in vitro fertilization.*

Certain genetic techniques have been applied to one of the greatest heartaches of childbirth. Many couples produce children born with severe and sometimes life-threatening deformities. Some couples have a genetic predisposition to produce severely disfigured kids, some so mutated that they do not survive long inside the womb. Others produce children who will carry with them diseases that will prematurely end their lives. Ideally, some couples would like to have a way of prescreening embryos to see which ones would have deleterious traits and which ones would not. The problem lies in assessing which embryo would display a certain trait and which one would not.

The selection problem has been surmounted by in vitro fertilization. The children derived from this technique have been misnamed test-tube babies. It would be more accurate to call them petri-dish embryos. In vitro fertilization is basically a way to simulate the fertilizing effects of sexual intercourse. The potential mother is first treated with fertility drugs to raise the number of eggs she will put through ovulation. These eggs are then harvested (that's the formal term) and eventually placed in a petri dish filled with fluid. The donor sperm are then harvested and dripped onto the eggs in the dish. These sperm fertilize the eggs, and the embryos begin to develop.

Typically twelve to twenty eggs are fertilized in this fashion and most placed into deep freeze. Four fertilized eggs are usually implanted into the mother's uterus with the hope that one will take. The only laboratory intervention involved in the creation of these pregnancies is the events surrounding fertilization.

Glass Distinctions

Numerous techniques have been developed to assess the genetic composition of an embryo. These techniques can accurately predict if a certain fertilized egg will grow up with a particular mutation. In vitro fertilization provides the opportunity to examine many candidate embryos before implantation. Those eggs without the bad trait can be candidates for implantation. Those with the bad trait can simply be disposed of.

In addition to genetic screening, certain cell culture techniques have been employed to keep human fertilized eggs alive for a period of time. These human eggs will begin to develop as if they were inside an adult female. They grow in a manner remarkably similar to the cultured animal embryos mentioned previously. Thus, a human embryo may grow into the *blastocyst stage* of development. The blastocyst is a complex multicellular entity that will be more fully explained in future chapters. An egg that develops in a human body will reach the blastocyst stage about four to six days after conception.

Because these cells can be grown for a period of time in culture, experiments can be performed on their viability if certain tissues are removed. If an embryo was going to be genetically assessed for defects, it could be grown to the eight-cell stage. Then a number of its cells would be removed and examined for the presence of certain defects. Could these embryos that had some of their cells removed develop into normal human babies? That is, could part of a human embryo complete the developmental process and, like animal cells, develop into a whole organism?

An experiment was done to determine whether humans would survive such a genetic assessment. The researchers decided to screen for *sex-linked developmental disorders,* specific to males. Over two hundred genetic disease mutations affect human males. These disorders are called *X-linked diseases* and include X-linked mental retardation, Lesch-Nyhan syndrome, and Duchenne muscular dystrophy. Some couples are at great risk for creating male children who will have one of these diseases. Such afflictions will not occur if the children are female. In this experiment, couples with a predisposition for these disorders agreed to undergo in vitro fertilization and create a number of embryos. Those embryos were allowed to grow in culture until they had reached the eight-cell stage. They would then be assessed for gender. Any males would be discarded; any females would be implanted into the patient's uterus.

A small drop of acid was applied to the surface of the growing eight-celled embryo with an extremely tiny needle, which created a temporary hole. An even tinier vacuum cleaner was inserted into the hole, and two of the eight cells in the embryo were sucked out. This technique removed 25 percent of the genetic information available to the developing embryo. These two cells were examined to determine whether the embryo was a genetic male or a genetic female. This determination process destroyed the two cells that had been isolated. Once a female had been found, the researchers returned to the six remaining cells and implanted them into the patient's uterus. Amazingly, those who became pregnant created normal healthy little girls. Just like the animal studies, this extreme genetic perturbation did little to stop the developmental process from occurring.

These results suggest that human cells have a large degree of manipulative flexibility reminiscent of the animal experiments. Since this genetic screening experiment was successful, an obvious line of experiments would be to determine if certain embryological manipulations performed routinely in animals will also work in humans. As more countries obtain permission to perform experiments on human embryos, these questions will probably be answered. And for better or worse, the formal possibility that viable fetuses can be constructed from bits of human embryos will be entertained. And probably be determined.

Gene Spirited

Human beings can do some absolutely amazing things. We have all heard of the mathematical wizards who can compute the answer to 14,325.781 X 1,636.459 more quickly than the numbers can be punched into a calculator. When they are asked how they did it, they often say something equally enigmatic like, "It was obvious. You just multiply 2,456 by 9,545 and then add 0.4207." The same phenomenon is true in art. Michelangelo's purported description of how to carve an adult human figure from a large block of marble was said to be, "It's easy. You just take away everything that doesn't look like a human being."

For most of us, these explanations seem more like excuses. The creative processes are so mysterious that those who possess them appear almost inhuman. And yet, these processes are strangely familiar because we relate so clearly to their owners; after all, humans perform these feats. Some might like to possess such genius if only it were possible. The situation is like that of a banker who hands us a bag of coal and says, "I have just given you a bag of diamonds worth an untold fortune. All you have to do is squeeze."

The deeply intricate processes that molecular biology is revealing to the human experience have similar frustrating aspects. Many tasks that genes and chromosomes perform are so fantastic that our attempts to explain them seem like trivial excuses. So little is known that we end up awarding Nobel Prizes to people just for framing the right questions. Any answers we receive from our explorations usually give birth to yet more questions; thus, the end result of most discoveries is that we simply find new undefined territories to explore. If the overall task of research is to reduce the number of answerable questions, science is a miserable failure.

Our task in this section is to address one of these mysterious biological questions. It can be phrased as something of a problem. All the genetic information necessary to describe a complete human being resides within each cell of your body (with the exception of certain cells like red blood cells). Yet, only sperm and eggs, which possess exactly half that information, are capable of making a human being. What is the difference between eggs and sperm and the rest of the cells in your body? What reproductive secrets authorize sex cells to create human beings and bar the others from following suit? Is it possible that these secrets could be unlocked from the vaults of a typical garden-variety nucleus? This section describes some progress that has been made in divining the reproductive secrets within cells, germ or otherwise. To accomplish this, we must examine the complex biology of nonreproductive gene regulation and human tumors.

And frog legs.

A Nose Is a Nose

Let's first examine principles about the genetic content of human beings. The human body works a lot like a very crowded city, one with between 20 trillion and 60 trillion cellular citizens. In previous chapters, we have seen that these cells perform specific functions like individual members of a community. Stomach cells make acid; cardiac cells beat in rhythms; certain skin cells make fingernails. Each cell has been genetically programmed to perform its specialized function within its cellular environment. Changing that cellular programming once it has been established can be extremely difficult. Thus, it is practically impossible to make a brain cell beat like a heart cell. Or get a cheek cell to grow hair like an outer skin cell. This particular degree of specialization is called *biological differentiation.*

These specialized cells possess identical copies of genes and chromosomes. An amazing thing about your body is the amount of genetic information within every cell. Each cell has all the genetic information necessary to completely describe a new you, a total of forty-six chromosomes in each cell. Buried within the cells of your armpit are the complete plans for an entire brain, heart, stomach, sperm or egg, skeleton, thigh muscle, and so on. But not just any brain or heart or thigh muscle. Your brain, your heart, in fact, all of your tissues. The cells in your cheek contain all that information, too. So do the cells in your stomach. And your scalp, your eyes, and your big toe. Your body has given the plans for your total genetic reconstruction to every cell that can handle it, which is practically all of them.

This fact leads to an intriguing series of questions. If all of the information necessary to genetically describe a complete human being is in each cell, why is a cheek cell always a cheek cell? Why do the cells in your liver not grow fingernails? Why do the cells in your big toe not grow kidneys? All the information is there. So are the cellular support mechanisms like nutrient transports and energy sources. A single brain controls a multitude of differentiated tissues, and a single blood system feeds the trillions of cells. Yet even if you wound your big toe and destroy some of those cells, you will grow back not beating heart cells but more toe cells.

The techniques of molecular biology are beginning to shed light on these and related questions. The answer is somewhat startling. The primary reason only fingernail genes are active in fingernail cells is that the other genes are silent. Asleep. They have been deliberately turned off. This turning-off capability is no small trick. From 100,000 to 150,000 genes are capable of being activated in a single cell. But in a stomach cell, the genes that would make a big toe, a kidney, an eye lens, or a human baby are silent. In your eye, the stomach-specific genes are turned off, and the eye-specific genes are activated. This on-if-you-need-it, off-if-you-don't pattern is found in all tissues of the human body.

Scientists are looking for ways to wake up these sleeping "Rip van Winkle" genes. Research has shown that genes in certain cells remain silent for extremely complex reasons. For example, proteins sit on some genes and actively repress them. Some genes are decorated with certain tiny chemicals that prevent activation. The three-dimensional way the chromosome is crammed into the nucleus may be important in keeping some genes active and some comatose. Some genes are silent for a period of time and then start blinking on and off like Christmas lights. Some are active only when the cells are young; some only when the cells are old. Understanding what turns genes on and off is one of the hottest areas of molecular biological research today.

So, the state of cellular specialization is a matter of genetic silence. Does this mean that creating tissues or entire organs is a matter of turning on a battery of specialized genes? Or activating not-so-specialized genes in the correct order? Does a fertilized egg turn into a complex adult because certain genes are turned on in a specific order? The answers to these questions lie within the genetics of an individual's developmental program. I have used *developmental program* rather loosely in preceding pages. To convey an understanding of the molecular basis of embryonic growth, I must now define it in more specific terms.

No Cause for Alarm

Two sets of biochemistries had to exist for a fertilized egg to transform us into something more complex than a couple of cells. First, all the raw genetic information needed to create a human had to be resident within the egg's watery borders. Second, the fertilized egg needed to know how to utilize the information to get itself to adulthood. The researcher's task has been to determine how such a tiny entity achieved such wisdom. Part of the answer is that the embryo follows a molecular owner's manual buried deep within its genetic code. This manual is the real developmental program, which tells the embryo what to do at certain stages of development, sort of like an order of worship in a church bulletin. As the embryo gets older, complex tissues unfold in an increasingly specific and sequential order, a process that reveals more pages of the developmental program. At each stage, the embryo looks to this manual and responds to its specific molecular instructions.

Who issues the orders? What is the nature of the instruction set? How does it read those instructions? We are slowly beginning to understand that a genetic chain of command is in developing embryos. We have found that some genes act like alarm clocks for other genes. That is, certain genes wake up other genes that had previously been silent. Some newly awakened genes are themselves alarm clocks. Their task, once awakened, is to perform certain functions and then turn on yet other sets of genes. These, in turn, perform their functions and then wake up other genes, and the beat goes on.

We soon realized that the turning on and off of genes was the mechanism embryos employed to construct themselves. Each new set of awakened genes was like turning a new page in the instruction manual. When one set of genes was activated, it barked certain genetic instructions to the next group of genes. Those genes performed those instructions, perhaps like "make a new liver cell here" or "form the beginnings of a spinal column now." When they finished, they turned on the next battery of genes so that the process could continue.

Eventually, an incredible symphony of genetic awakenings was observed. Each awakening formed the next part of the organism until, in a shattering climax, a baby was born.

A Lock and a Hard Place

Because a baby starts out as two cells, and its genes follow an established hierarchy, a master gene or series of genes must get the whole thing started. The scientific hunt for these master switches that control the beginnings of these developmental processes is at a fever pitch. Because the genetic code is in every cell, we already know where, and how, to get to the structural information. Deciphering the order of the grand genetic awakening is more elusive. Which genes are turned on first? Which next?

What growth factors turn them all on? If the presence of the structural genes can be defined as the presence of a safe, the order of their activation can be defined as the combination that unlocks it. Because those structural genes are in every cell, copies of the safe reside in all tissues. If we can learn how to pick the biochemical lock that opens the safe, we will be able to control the developmental program.

Many animals have already discovered the combination to their own development. And we are slowly beginning to understand how they found out. This biochemical safecracking has been observed in organisms as tiny as insects and as large as palm trees. Understanding the developmental mechanisms of these creatures may tell us how to unlock our own.

Growing Pains

The insect world is full of citizens with at least a partial genetic memory of their youth. Insects have a hardened exoskeleton that surrounds a soft body. To grow a new outer skeleton, they have to cast off the old one, like one might change clothes. This process is called *molting.* As long as an insect continues to molt, it can regenerate virtually any appendage that hangs off its body. That is, if you cut off the legs of most insects, they can regrow them. You can cut some insects in half, and they will grow back their posterior ends. You can decapitate one organism and it will grow back another head on the stump of the old one.

Higher organisms can partially regenerate certain tissues. The leg of a starfish can be amputated, and the starfish will grow back another in the right place. I once tried to capture a lizard that had a bright blue tail. I grabbed its tail only to find it was fully detachable. The lizard ran away in the grass, and I was left with a lifeless blue tail. I later learned that these lizards have the ability to grow back as many tails as there are little boys to grab them.

Insects and lizards demonstrate that the developmental program can be reinitiated for specific body parts that undergo trauma. They do not need the presence of a fertilized egg to tell them how to do it. Are there more dramatic examples of body regeneration from normal nongerm cell tissues? Is it possible that organisms can grow entirely new selves if just a few hunks of nonsex cell tissue are left? The answer to those questions is yes. The two examples we will discuss are sea cucumbers and, of all things, house plants.

Total Recall

Sea cucumbers are ocean-dwelling invertebrates that look more like pickles than cucumbers. In times of danger, overcrowding, or environmental stress, their bodies can jump ship. They have the disgusting ability to vomit out their lungs, their digestive organs, their gonads, virtually their entire internal anatomy, through a series of forceful contractions. The only things left (depending on the species) are the rectum and parts of the body shell. This process is known as *evisceration.*

The sea cucumber can completely regenerate itself. It needs only an intact rectum to rebuild an entirely new body. There are no fertilized eggs around and no sperm; all the sex organs are usually jettisoned with the evisceration. This rectal tissue, though normally reduced to humble service, has a long genetic memory. Some form of developmental program is incited by those cells in the complete absence of any reproductive tissue. A complete sea cucumber is usually reformed about three weeks after evisceration has occurred.

Even though sea cucumbers have amazing abilities to recover from self-inflicted injury, the champion regenerators belong to plants. Botanists have placed small chunks of leaf cells into specialized cultures. These cells are allowed to grow into a large mass of disorganized tissue that looks like mounds of mashed potatoes. By adding certain chemicals to the culture, they can get roots to form out of the bottom of this mass. By adding still other chemicals, they can coax shoots to form out of the top. An entire plant can be regenerated. It is an amazing developmental feat.

The mammalian equivalent would be to cut out a belly button and place it into a dish. Then through the addition of certain chemicals, a leg or a chest or both would be induced to begin growing out of the culture. That is nonsense to mammals but is routine to plants. Plants are thus the regeneration heroes. In this case, the ordered gene switching is readily available through unsophisticated environmental commands; the plant needs very little input to trigger the developmental program in its entirety.

A Genetic Toehold

Such development is apparent in several members of the animal kingdom. Groups of cells capable of regenerating entire organisms are said to be *totipotent.* So far, we have discussed inciting developmental programs only from totipotent cellular groups. However, I have stated that enough genetic information exists within single nonsex cells to create embryos. Can the development of an entire organism proceed by utilizing the genetic information in, say, a toe cell? The experiment would be to place that toe cell under the best creature-generating circumstances possible. If it grew part of the way to being an embryo, the experiment could be called a success. This experiment has been done (although not without some controversy about the results), and its results give us reason to reconsider what conception is all about.

This is what happened. Learning a lesson from the sheep people, researchers removed the nucleus of a normal unfertilized frog egg. A toe cell from the frog's front leg (really the foot web) was placed into culture and allowed to grow. After a period of time, a nucleus from a cell in this culture was extracted and placed into the enucleated egg. The combination of cells was incubated for another period of time.

To just about everyone's surprise, the egg developed into a tadpole. Except for an intermediate step, it was almost as if a normal egg had been fertilized by a normal sperm. The fully differentiated toe cell responded to cellular signals inside a genetically emasculated egg. The developmental program was incited and from that toe cell sprang nerves and a blood system and muscles and the ability to move. This fact is profound because the normal distance between a toe cell and a whole organism was thought to be practically infinite. In this experiment, that distance shrank to zero.

The ability for a toe cell to create an entire organism sets the stage to seek the developmental potential of all cells. The only genetic information supplied to the tadpole came from the leg of an adult frog. The biochemical safe, with all the information necessary to create a new frog, was there. No big deal. All frog legs have that potential. You do not see tadpoles growing off the ends of them, however. The combination that opened the lock on the safe was found in an unfertilized egg. There was no genetic information in that egg. The egg hadn't even touched a sperm. Yet, the chemicals in the burned-out egg united with the toe genes in its interior and created a complex and living clone. To be sure, there is some controversy about this result. Some researchers were skeptical of the low numbers of converted eggs; some were not sure if all the eggs had been properly enucleated. Exactly what had been achieved, in the minds of some, was an open question.

If it is true, the result is an incredible achievement. The idea that the genes holding the developmental keys exist in toe cells is beyond dispute, even if certain doubts remain. Regardless, it gives us reason to rethink what we believe about the developmental status of all cell types. And other species.

It is also an achievement that would be easy to overinterpret, especially when we want to understand human biology. We must remember the great differences in the ways animals use their constituent biochemicals. Most mammals don't eviscerate. As Anne Boleyn can testify, humans will not regenerate a head after it has been removed. You will never grow another finger by continually immersing your hand in water. In the toe cell experiment, the tadpole never developed into a frog. There is such a tremendous amount to learn that we probably do not know enough to project to other species.

Nonetheless, could something similar to this experiment happen to human cells? The answers to such experiments would shed light on extremely critical definitions of human life. Those stakes are heavy enough that the implications of any reproductive experiment, overinterpreted or not, must be taken seriously. The toe cell experiment demonstrates tremendous flexibility in the stimulation of the vertebrate developmental program. There are some hints that the developmental program may be flexible in humans as well. In which case the answer to the above question may be yes, eventually. This hint does not come from a wayward research enterprise. It comes from a very natural, and very bizarre, form of tumor.

A Weird Sense of Tumor

A teratoma is a cancer that arises primarily from germ cells. It can be found as a tumor of the ovaries or the testicles. It can be found on the side of intestines. It can be found in babies whose spinal cord never finished forming. It's not an ordinary tumor, however. It was described a little bit in chapter 1. As you recall, if a teratoma tumor is removed from a patient and dissected, a grotesque sight is observed. It is the only tumor in which human teeth can be found. So can lens tissue of the eyes. Entire fingers can be in the tumorous mass. So can miscellaneous bone fragments, lung tissue, and human skin—complete with hair.

It almost looks as if the mass of tissue was at one time genetically trying to become a human being. Somehow it didn't get the order of the gene awakening correct or didn't have the right three-dimensional cues or correct secondary tissue support. It appears instead to have collapsed into an inhuman, disorganized poor substitute. This is not the fusion of an egg or sperm cell. It is not necessarily a blastomere gone awry. It is simply a tumor.

The biology of such tissue is under intense investigation. Parts of the developmental program appear to have been initiated in these human tissues without the benefit of fertilization. These tissues have been placed into culture and experiments have been performed on them. The fruit of these experiments has been the isolation of a number of cells that can be converted into differentiating masses of tissue. Similar to the plant experiment, adding different chemicals incites different tissues to develop.

This result is exciting in that it mimics aspects of human development in a dish. Valuable alarm clock genes may be harvested from such tissue. The instructions that such genes carry with them before they wake up the next set may be deciphered in these genes. Perhaps even certain master signals will be discerned. This tissue has provided the experimental material with which to study human development without experimenting on people. It will be the headwaters at which the developmental program of human beings, the genetic instruction sheet that turns two tiny cells into babies, may be isolated. Teratoma tumors might, in an odd sort of way, already be showing us how this is accomplished.

Whether we are discussing parthenogenesis or frogs from feet, these experiments reveal a central biological fact. Conception as a milestone is not the only mechanism that exists to ignite the genetic beginnings of life. It is a process as flexible as the ways sperm and eggs get together and as changeable as the genders that escort them. The real question concerns the molecular switches that kick start a developmental sequence leading to an embryo; it is not the joining of two unlike cells. As has been discussed, you can bypass conception under certain conditions if you can get to the switches. Remarkably, God has designed it so that the biology lends itself to such study. We have responded with intensity, attempting to squeeze the coals of our data to see what it is made of and what we are made of. Whether we are strong enough to make diamonds from our efforts remains to be seen.

Fluid Dynamics

The disturbing conclusion from most of the animal experiments is that fertilization is just one way to create functional embryos. Other processes, utilizing lizard eggs, split cow embryos, and toe cells, are also available. If those processes were confined to animals, we might be more fascinated than fearful of the work. The problem is that we see hints of similar processes in human cells. The knee-jerk reaction to those hints is to apply all of the results from animals to human cells and then react with hesitancy. In reality, the animal research doesn't say diddly about what goes on in humans. It can only offer suggestions about human processes. But even as suggestions, the experiments can give us reason to pause.

As the knowledge base grows, the issues surrounding human development will become increasingly complicated. The purpose of this chapter has been to define certain aspects of that complexity. And perhaps to define areas of confusion. We are used to thinking about biological processes in monolithic terms. The problem is that they are not always solid, fixed, every-living-thing-does-it-this-way processes. These processes are fluid.

To understand this fluidity, we have had to go to the source, the genders that create the eggs and sperm. We have seen that organisms can be both genders or neither or switch between them, depending on their particular internal genetic architecture. These facts point out that sexual determinations are not monolithic biological absolutes. Gender is a function of certain genes in biological organisms. These genes can be incited to create certain cells, or they cannot. The cells created by these genders will confer on them certain reproductive capabilities. There is nothing set in stone about it.

This fluidity sets the stage for the next point: you do not always need two genders to make a fully functional embryo. Parthenogenesis demonstrates that certain genetic mechanisms bypass normal fertilization. In many instances, these mechanisms are as reproductively successful as their sexual counterparts. Conception by sperm and egg is not the only way to create an embryo; it is one way to create an embryo. Inciting the genes to kick start the developmental program in a cell is the most important event in the making of an embryo. Parthenogenesis tells us that inciting this developmental program is not the exclusive creation of a sperm/egg joining. The program resides in the egg without the sperm's input. That parthenogenesis can be induced says this is true even in creatures that normally sexually reproduce.

Parts Is Parts

The fact that conception is not needed to make embryos demonstrates the power of the genes in the developmental program. However, we are not just working on events that bypass conception. We are also learning to keep the resultant embryos alive in petri dishes. These technologies have shown us that the developmental program exists in cells other than sperm or eggs. You can cut an embryo in half and place each half into a surrogate mother. Each half can examine what it is missing and replace the absent cells; two genetically identical embryos are created. Presumably, you can keep cutting them in half as they develop and so obtain great numbers of identical organisms.

You don't really need to cut an embryo in half. Just take one cell from a certain stage of a developing embryo and put that in a mother. A new one will form. You don't really need a cell. A nucleus from a fertilized egg will do nicely.

The conclusions of these experiments are the same as those concerning parthenogenesis. Not just the sperm and the egg carry all the developmental information needed to create a finite organism. The wealth has been spread around to whole embryos, parts of embryos, and nuclei. The common denominator is the genetic developmental program. The joining of the sperm and egg is not the key to making an embryo. Once again, the key is the ignition of the developmental program, however it is started.

Finally, we are beginning to understand the molecular nature of this developmental program. Some animals retain portions of the developmental program throughout their lives, which permits them to regenerate sections of themselves when the need arises. We are beginning to understand that this developmental program consists of a hierarchy of increasing genetic awareness. Preexisting genes are turned on and off in a prescribed sequence that eventually results in the creation of an organism. Disturbingly, we have found that the structural genetic information has been bequeathed not just to specialized sex cells but to all cells. Thus, the creation of functional embryos may not require anything more than the intact genetic information from a few bits of skin. These facts once again point to a familiar conclusion. The joining of two specialized cells doesn't define the beginnings of life. The developmental program that this joining incites carries the real definitions.

I used to think we could ignore these issues because of their irrelevance to the real world of research. What science is actually capable of doing and what people fear it is capable of doing are often two different things. People will react with alarm to an experiment done with animals as if the human equivalent were just around the corner. And the over-interpretation of some animal experiments hasn't helped to temper that reaction. Researchers have become overenthusiastic about what a result means to the species studied, no projection onto humans required.

But I do not ignore these issues anymore. Each fact we have talked about has at least representative experimental momentum in human cells. The legislation passed under the recommendations of the Warnock Commission and the permission already granted in other countries are real events. Such legislation will allow many experiments originally performed on animal embryos to be tried on human ones. We may know if we can endlessly clone humans like we can endlessly clone cattle. We may learn how to carry a parthenogenetically derived fetus to term. We may be able to create the tissues and perform the experiments on human toe cells to see if we can turn them into embryos. Most important, we will become increasingly familiar with the incredible developmental program that turns tiny cells into complex adults.

And the way we think about the beginnings of human life will be changed forever.

It is possible that molecular biology has introduced to us a construction site we may not want to visit. Do you hear any dogs barking?

Chapter 7

If Looks Could Thrill

When Mr. Job picked Scotty and me up after a trip to the aquarium, we thought we had died and gone to heaven. We were filled with visions of those giant fish, floating in aquariums the size of small automobiles. Mr. Job, a tough marine sergeant who doubled in the even tougher role as Scotty's dad, was silent in their station wagon. That's because we weren't. We were busy making plans that would allow us to duplicate at home the experience we had just witnessed. Even if we could achieve only the appearance of those giant aquariums, our enthusiasm would be satisfied. We didn't realize that creating the appearance wasn't the best way to duplicate what we saw. We were to find out how much trouble substitution could get us into.

By the time the station wagon pulled into Scotty's garage, we had already concocted a plan. It would involve the Jobs' small family aquarium and some fish that Mr. Job had caught and put in the freezer the weekend before. Our plan was based upon the sound scientific notion that frozen fish were the biological equivalent of sleeping fish. Knowing that grown-ups never appreciate the value of simulation, we waited until Scotty's parents were downstairs watching the evening news. We then quietly took the fish out of the freezer, unwrapped them, placed them in the aquarium, and waited for them to magically spring to life.

There was no luck. The fish sank to the bottom. They could not be prodded into swimming. They could not be poked into swimming. They just lay there, bewildered angelfish sniffing at the frozen invaders. We began to wonder what went wrong. Perhaps the frozen fish didn't look alive because the home tank was too small. Maybe they would appear more alive if we moved them to a larger enclosure.

That's when Scotty got a bright idea. He reasoned that the biggest enclosed area that would appear like the aquarium was their station wagon. So we took the fish out of the tank, put them in the front seat of the car, and rolled up all the windows except for a little space on the driver's side. We got the garden hose, dragged it over to the car, inserted the hose into the little space in the window, and turned it on full blast.

You know, for a while, the plan actually worked. We could see the fish seeming to float to the surface. Though it wasn't quite Sea World and the fish weren't really floating, the appearance was good enough for us. We were so proud of our achievement that Scotty ran back to get his dad so father and son could admire the new aquarium. A minute later, to my surprise, his dad came roaring out of the basement ahead of Scotty. He emitted a sound that can be described only as a cross between a stockbroker who has just lost millions and a dying wolf. It was the only time I ever heard Mr. Job comment on the sexual history of his wife and my mother. He, like most parents, did not understand that responsibility genes are generally not active in nine-year-old heads. Later that evening, Scotty and I got the biggest lickings of our little lives. We learned that the effect of simulations on people can be deceiving. And that one must always be careful in creating and judging things based solely on appearances. Or at least be discriminating in who you display your creativity in front of.

I have learned a lot about appearances since that time. Mostly, I mistrust them. There is a danger in relying on how things look to tell us how things are. From politics to advertisements, appearances that are substituted for the real thing devalue the real thing. Or worse, make the real thing irrelevant. The folly comes when we rely on appearances to tell us things about our world. This can lead to disastrous consequences, like poor Mr. Job's station wagon.

The subject of this chapter is a certain biological appearance. A human appearance. Some of us would like to settle the question about what is human and what is tissue simply by appearance. This approach is the opposite of that of Descartes. The view holds that human beings exist not because they can think but because they can tan. If the tissues appear to be like a human being, they must be a human being; conversely, if the tissues appear to be less like a human being, they must be inhuman.

We can examine the roots of such physical judgment by contemplating the time frame of a pregnancy. We do not have a large moral momentum to save a sperm or an egg because of its potential to create human life. Neither appears human to us. We do have a large moral momentum to save a healthy baby after it is born. Such a collection of cells appears human to us. Something occurs between the time of productive intercourse and the time just after delivery to change our minds. What happens?

A change in the appearance of these cells happens. Two not-so-human-being-looking cells turn into several billion that collectively look exactly like a human being.

Perhaps a sports illustration is useful here. We generally do not define a golf game by the presence of a ball, a club, and a willing participant. They must work together in an organized fashion for a golf game to happen. It is defined not only as the presence of tools but also as the performance of those tools. Is the same true for certain reproductive questions? We generally do not define third-party life by the presence of sperm, egg, and two willing participants. They must work together in an organized fashion for a baby to be produced.

Since we do not have a problem with unjoined germ cells, only two biological issues can be examined in human origins: conception and subsequent development. We have already discussed in some detail the process and ambiguities of conception. Now we must deal with the biological consequences of development. If we were on a green, we would have to decide when golf tools convert to golf games. Definitions could come by the presence of a competitor or a club's hitting the ball or the ball's flying in the air or all of these or none. Somewhere we have to define a marker that tells us when the game commences. Similar questions apply to the embryo. We must examine the process of embryonic growth to see if there is a biological milestone that will unequivocally tell us when human tissues convert to human beings. Summarily stated, the biological question is this:

Can the physical appearance of a human embryo be grounds for ascribing humanity to it?

To fully appreciate this question, we must understand the biology of development. The purpose of this chapter is to examine how an embryo grows, from its silent two-cell preembryo stage through its crying, cooing two-billion cell fetal stage. The scope is necessarily brief. The genetic descriptions will have substantially changed by the time you read this anyway. However, without this background, we cannot begin to answer the question of events that occur postconception. With it, we can at least create the appearance that we can.

Leave It to Cleavage

Time has often been defined as "a referee that keeps everything from happening all at once." Such physical mediation has produced a biological convenience. It has allowed us to view the development of human embryos as a sequential series of well-ordered events rather than as a sudden burst of spontaneous creation. We have divided these events into a series of stages. Often called the *cleavage* or *preimplantation stage,* the first stage in human development covers the initial ten days after conception. The goal at this juncture of life is to grow like a weed and set up an embryonic tent that will support this growth for the next nine months. The preembryo starts out this complex series of events as a two-celled, mobile free agent. It ends this stage as an intricate collection of cells that has claimed squatter's rights on the side of the mother's uterus (see fig. 7.1).

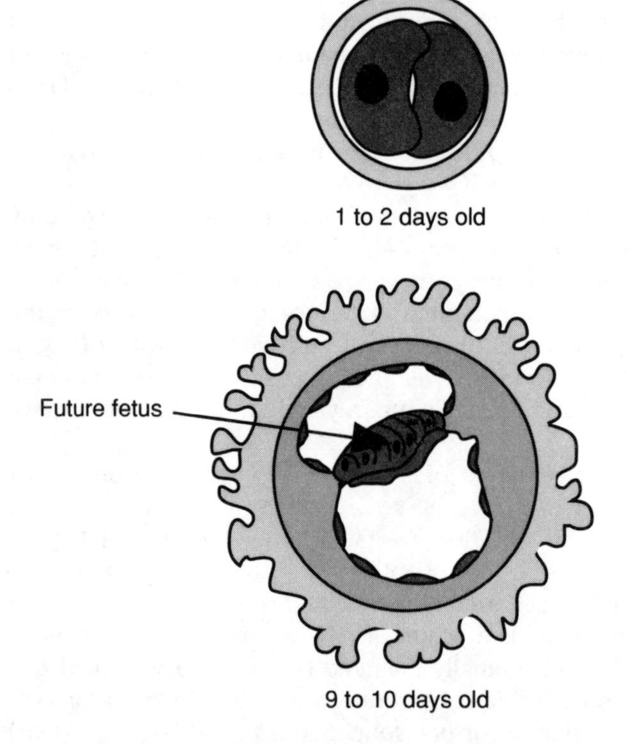

Fig. 7.1. **The first weeks of life. The preembryo shown at the top is two cells in size. The embryo shown at the bottom, already implanted into the side of the uterus, is many cells in size.**

Let's pick up on the events happening in the woman's uterus described in chapter 4. The last time we left our preembryonic hero, the sperm and egg pronuclei had just fused. Now what happens? The paradoxical function of this pair is to split as soon as they unite and make more cells. The two cells transform into four cells, and the four make eight cells, and so on (see fig. 7.2). Usually. They do not always cleave at the same time, however. Three of them might divide (making six), and the fourth might decide to take a vacation. An intermediate seven-cell stage is possible on its way to becoming an eight-celled preembryo. You probably recall from the last chapter that these cells are known as blastomeres regardless of their initial numbers.

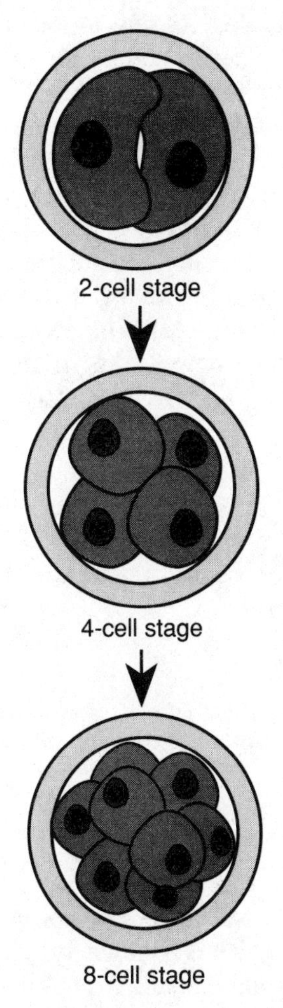

Fig. 7.2. **Stages of growth after conception has occurred. By the time the eight-cell stage is seen, about two days have passed.**

Blastomeres have the ability to undergo mitosis and walk at the same time. Eggs are normally fertilized in the Fallopian tubes. Those tubes contain tiny hairs that move in a coordinated motion, a lot like fronds of seaweed swaying in an ocean current. Called *cilia,* these hairs provide a gentle motion that maneuvers the fertilized egg toward the uterus while these early divisions occur. When we discuss preimplantation development, we do so with the preembryo in a constant state of motion.

The fertilized egg at the eight-cell stage is a loose arrangement of blastomeres tucked inside what looks like a translucent bubble. There is plenty of space between them. A dramatic internal change in this spacing takes place soon after this stage of development has been reached, however. These eight cells suddenly huddle together as if they decided to become an incomplete football team. Some cells become totally enclosed by their colleagues. The huddling produces maximum cell-cell contact.

The outer cells form little connections (*tight junctions*) between each other. In so doing, they seal off the internal cells from the rest of the world. The internal cells then drill little holes into their surfaces to form molecular bridges with their outer neighbors. This allows for the free exchange of important molecules among all the cells, whether they are on the outside or the inside. The entire huddling and building process is called *compaction*: a fully communicating multilevel creature is formed, possessing both external and internal cellular anatomies (see fig. 7.3).

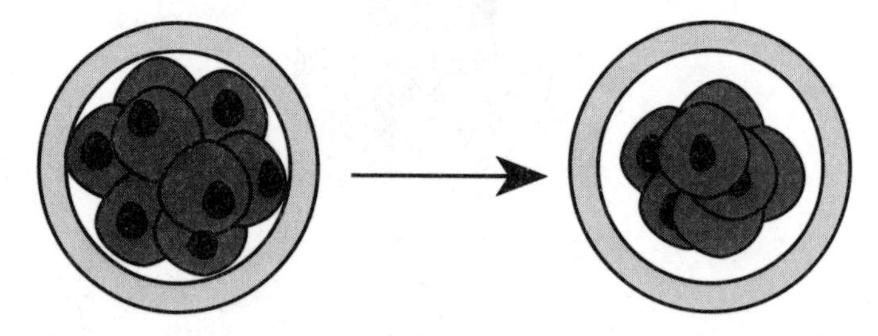

Fig. 7.3. **At the eight-cell stage, the developing cells undergo the process of compaction.**

Divided We Fall

After the cells are fully connected, the next task is to continue their divisions beyond the eight-cell stage. Division must take place while the cells are fully connected, which is a lot like doubling the number of beads on a necklace without cutting the string that holds them together. As you might suspect, scientists have given this connected-yet-dividing preembryo another name. It is a *morula,* a structure composed of twelve to sixteen cells. Most of these cells are on the outside; by the time the preembryo reaches the thirty-two-cell stage, only a few cells are on the inside. The rest of the interior is a giant, empty cavity. A typical morula looks something like a beach ball.

At the morula stage, the tiny preembryo falls out of the Fallopian tube and has its first close encounter with the vast expanse of the uterus. Also at this time, the morula starts secreting fluid into its empty inner cavity, now giving it the appearance of a beach ball filled with water. Those internal cells start positioning themselves onto one side of the inner wall, which creates an asymmetrical heaviness inside the developing preembryo. So many changes have occurred to the morula that giving it another name proved irresistible to scientists. The embryo is now a *blastocyst* (see fig. 7.4).

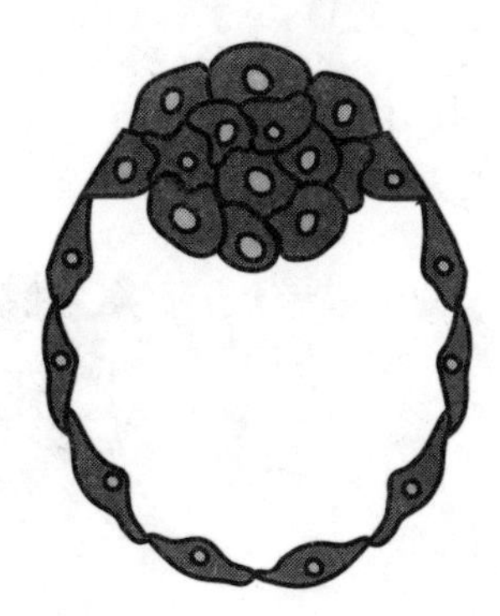

Fig. 7.4. **Cutaway view of a late blastocyst (about sixty cells). Note the aggregation of cells near the top of the structure.**

The Complexities of Implantation

As this blastocyst gets bigger, major distinctions can be observed between the outer cell lining and those asymmetrically placed inner cells. They become as different in their physical appearance as Mother Teresa is from Bette Midler. They begin to perform different functions. Even their internal biochemistries are altered. The developmental program has begun to awaken one set of genes in those outer cells and a somewhat different set of genes in those inner cells. A genetic division of labor has started, an unfair division in that outer cells outnumber inner ones. Although it doesn't seem like much, this division of labor is the all-time critical milestone in early human development. It is the first clear evidence that the developmental program has decided what will become a valuable baby and what will become a discardable tissue.

This decision is most easily observed by following the fate of the outer and inner cells of the blastocyst. The majority of these cells, the outer ones, will not become a human being. Ever. Not even part of a human being. These cells are destined to become part of the *placenta* (the *afterbirth*), a collection of tissues whose main function is to steal food from the mother and give it to her tiny passenger. These outer cells are called *trophoblasts* (see fig. 7.5).

If most cells of the blastocyst are good only for making afterbirth, how is a human being ever constructed? The inner cells, the ones clinging to

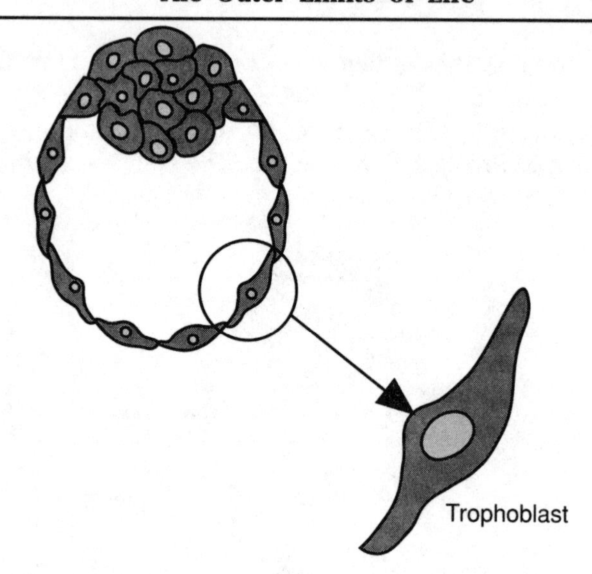

Fig. 7.5. **Trophoblasts are the outer cells of the blastocyst. These cells are destined to become the placenta, the membrane that helps in fetal nourishment. Trophoblasts, however, never become part of the fetus.**

the inner side of the blastocyst, will create the fetus. At this point, there are only a few of them. But within those few cells lies the biological equivalent of the big bang; the infinitely complex biochemistries that will drive an adult human being originate from the common wisdom of these cells. These inner cells are sensibly termed the *inner cell mass*. And the developmental program they carry will ensure the presence of more than a few cells very soon (see fig. 7.6).

The wandering blastocyst, with its division of labor firmly in place, is now free to settle down. The outer cells, the trophoblasts, start making a group of cells that are the molecular equivalent of an invading army. The soldiers in this army possess multiple nuclei within a single cell body, kind of like raisins in a large bowl of pudding. When the blastocyst encounters a uterine lining, this army partially slithers off the blastocyst and establishes a beachhead on the wall of the uterus. I say only partially because a good deal of it is still attached to the preembryo. The blastocyst becomes tethered to the conquered patch of uterine tissue. This tethering is part of a process called *implantation,* which generally occurs six days after conception and represents the end of the mobility of the blastocyst. The little passenger will now be tied to this position until it is born (see fig. 7.7).

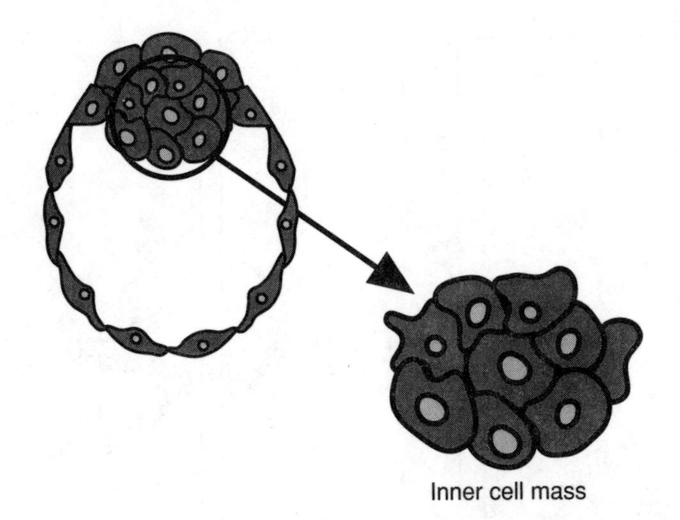

Fig. 7.6. **Asymmetrical division occurs on the inner side of the blastocyst. This division results in the formation of the inner cell mass. The inner cell mass possesses the cells from which a fetus will be made.**

The Yolk's on You

The embryo, firmly embedded in its uterine home, must now perform two tasks: (1) it must hook up to the mother's food and an oxygen source so that it can eat and breathe for nine months and (2) it must use that fuel to make a baby out of what is essentially a ball of complex cells.

Implantation assures that the preembryo will get food from the mother. How does this happen? As in human society, you can generally tell how people eat by looking at where they live. In most members of the animal kingdom there are two choices; the developing progeny either grow inside their mothers throughout pregnancy or are expelled and grow outside their mothers in the form of eggs. Animals that lay eggs do not have to worry, for the most part, about feeding their young in the early stages of growth. The egg is outfitted as one giant food supply house. The food supply is the yolk, configured so that the embryo can snack at will inside the egg as it develops and matures. The embryo is fed automatically, absorbing the proteins in the yolk as it forms its complex internal anatomy.

Human preembryos make yolk just like chickens do. Unfortunately, it is not enough to supply the developing fetus with dinner for its entire nine months. Human cells, like most mammals, have had to find another strategy in order to eat. The tactic, in a pattern to be repeated over and over again in life, is to look to mom for help. If it can connect to her

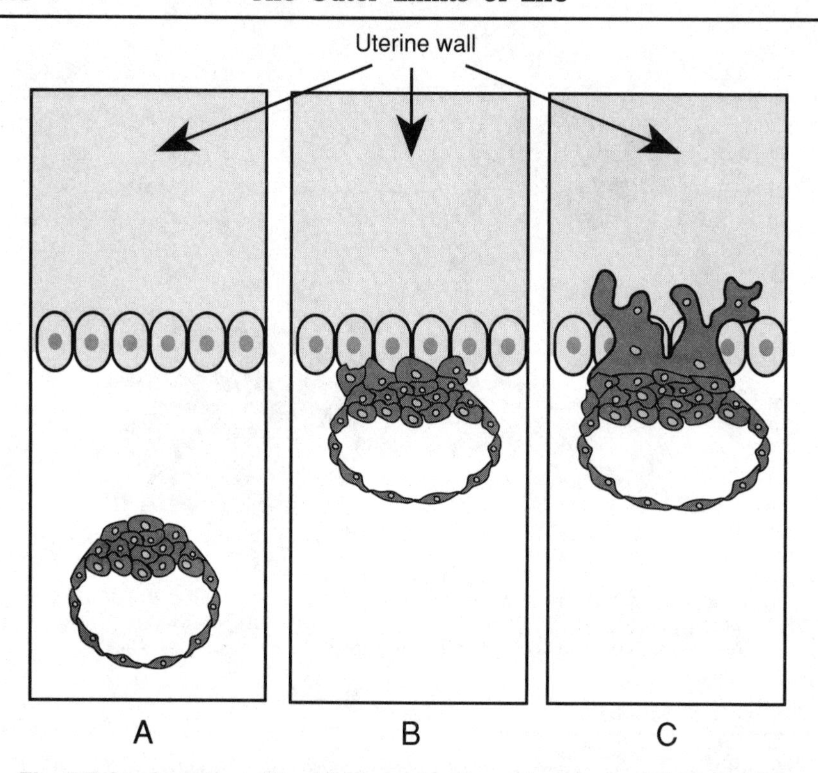

Fig. 7.7. **Implantation. The blastocyst enters the uterus (A) in the first week postconception and attaches to its lining (B). Multinucleated cells soon arise from the blastocyst and invade the uterine wall (C), permanently tethering the blastocyst to the side of the uterus.**

bloodstream, the preembryo can absorb nutrients and oxygen every time the mother eats and breathes. But it must find a suitable cellular linkage.

So how does the preembryo hook up? Is it like an RV at a campground that needs to connect certain pipes to itself in order to function? This hookup cannot be merely for food. As it matures, it will generate waste products. How will a fetus survive for half a year without turning the womb into something eligible for Superfund Cleanup money? Obviously a two-way exchange system is needed with its mother, both to give and to receive. When the preembryo implants into the uterus, this exchange system is the first thing it establishes.

The Sound of a Distant Plumber

The raisin pudding-like army plays a critical role in tethering the preembryo to the uterine wall. It also plays a critical role in hooking up supply lines to the mother. When those cells establish their beachhead, they also penetrate the inside of the uterus. Under the microscope, it looks just like a cellular invasion. The uterus, in response, sends blood vessels out to greet the wall of marauding embryonic tissue.

There the two will face each other and hold their positions in an uneasy truce. The preembryo, hanging out in the uterus, knows when its invading army has contacted the mother. It responds to this knowledge by doing what it does best: dividing like crazy. This dividing results in a stream of cells sent out to reinforce the raisin pudding army at its uterine beachhead. The newly recruited cells will eventually make the *umbilical cord,* which will ultimately be filled with blood vessels, structures that can soak up any nutrients encountered.

A kind of demilitarized zone is created between two opposing sides. The mother's blood vessels, carrying all the nourishment, are on one side of the uterus. The embryo's blood vessels, extending from its little body and ready to receive food and oxygen, are on the other side. In the middle are the raisin pudding cells. These cells will eventually turn into the sophisticated structure of the placenta, the nutritional go-between for the mother and her passenger until birth.

When fully established, the whole feeding system works something like this. Let's say the mother eats a pizza. If she is a typical human, she will also breathe between bites, depending on the quality of the pizza. Her body will turn that pizza into nutrients, which will then be placed into her circulatory system. The oxygen will also get into her blood by the more direct route of her lungs. Her circulatory system will siphon both nutrients and oxygen into the blood vessels surrounding her uterus. These vessels will, in turn, dump those molecules at the doorstep of the placenta, the organ originally made from the raisin pudding cells. The cells of the placenta will grab the nutrients and take them inside. It will toss them like a hot potato out the back door to the waiting arms of her little passenger's umbilical cord blood vessels. These vessels will slurp up the nutrients and give them to the baby, which can then use them to fuel its growth (see fig. 7.8).

The placenta, in addition to serving as a food broker, also provides a sewage disposal service. The embryo will eventually give off waste products, mostly urine and carbon dioxide. The cells of the placenta will immediately absorb the waste, internalize it, and deliver it to the waiting maternal blood vessels on the other side. The maternal vessels will absorb the waste, which pollutes her normal circulation. The mother will dispose of her passenger's waste in the same way she disposes of her own.

In this preembryo-driven arrangement, the circulatory systems of the mother and the child never really meet each other. The connection is established somewhat cooperatively but only at the molecular invitation of the preembryo. The developmental program of a fertilized egg carries information about how to construct a human baby and how to feed it.

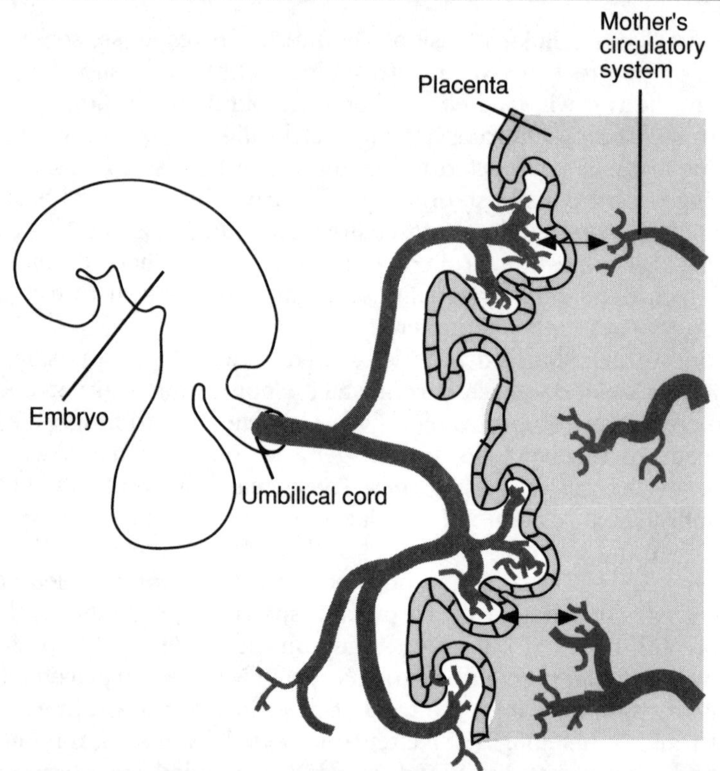

***Fig. 7.8.* Exchange of wastes and nutrients between mother and embryo. Shown here is part of the circulatory system of an embryo four weeks old. The tissue labeled placenta arose from the trophoblast cells (the outer cells of the blastocyst). The arrows indicate areas of exchange between mother and embryo.**

Genes are continually being activated and deactivated by the program to implant the preembryo and get it hooked up to mom. So that the preembryo may respond, the raisin pudding cells must constantly send genetic messages apprising the preembryo of its status. The same pudding cells can activate genes in maternal cells, primarily to get mom ready for pregnancy. Even though there is no formal connection between the circulatory systems, there is a tremendous amount of communication between the mother and her developing passenger.

Twinning Streak

The connection just described is relevant to pregnancies producing only one child, a situation true for eighty-nine out of ninety births. But it's not true for every pregnancy. In some situations the hookups are so varied that the mechanisms producing them change the way we define human life. This is the biology of multiple births. Before we finish our discussion, we need to talk about this twist of the developmental program. In the following examples, we will focus on the manufacture of twins, the biology of how two humans can originate from a single uterus.

Two types of twins can come from human beings. One type is called *fraternal* or, in scientific obscurese, *dizygotic*. The other is called *identical* or *monozygotic*. Fraternal twins are formed as the result of the mother's being generous with her menstrual cycle. She sheds two separate eggs, which upon intercourse are fertilized by two separate sperm. The resultant babies are not identical. They are no more related to each other than any other nontwin sibling in a family. They can be different sexes or female, casting an odd light as to why they're called fraternal. These biological duets are simply the same age (see fig. 7.9).

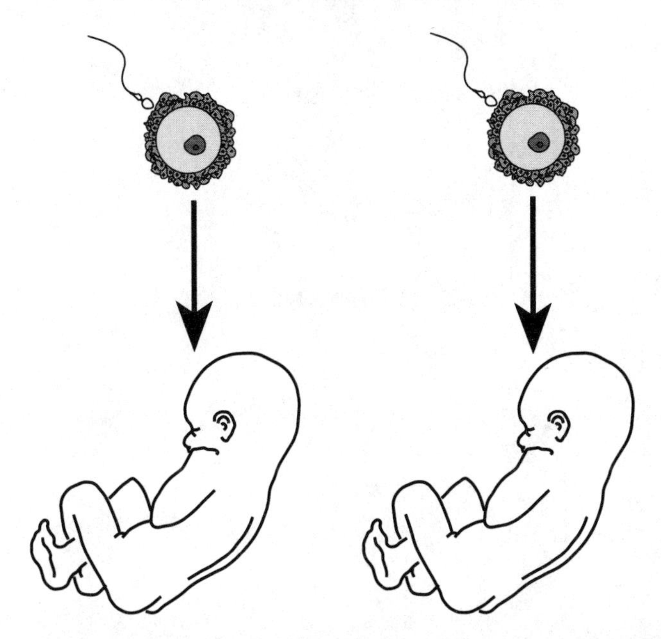

Fig. 7.9. **Fraternal (dizygotic) twins. These twins arise from the fertilization of two eggs, the result of a double ovulation. Although they are born at the same time, these twins have no more genetic identity than siblings arising from separate pregnancies.**

Identical twins, on the other hand, are just that, genetically identical. They are not formed as the result of two separate fertilizations. They are created as the result of a single fertilization (see fig. 7.10). The splitting event occurs after the embryo has already begun forming. This separation causes two genetically identical humans to be formed from a single fertilized egg. The same is true for identical triplets, quads, and quints. In these cases, the egg does not split into two; it splits into three or four or five. At conception, the exact number of preembryos that will be created cannot be determined.

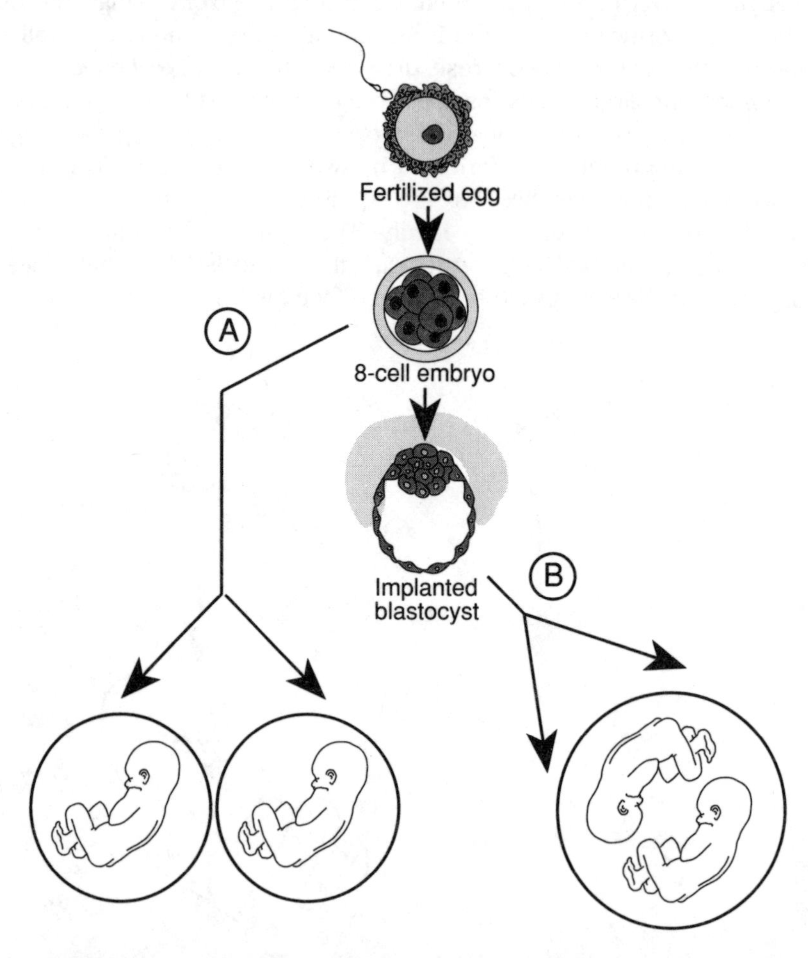

Fig. 7.10. **Identical (monozygotic) twins. These twins arise from a single fertilization event, followed by splitting of the preembryo. If the split occurs prior to implantation (A), two separate placentas will be formed. If the split occurs after implantation (B), one placenta will be formed with the fetuses sharing an umbilical cord. These twins have the same genetic information and, regardless of the split, will be physically identical.**

Like all preembryos, the fertilized eggs of twins must undergo implantation to survive. For fraternal twins, each blastocyst finds its own spot in the uterus and digs in for the nine-month haul. Separate placentas and umbilical cords are created, and the mother gives birth to two genetically distinct individuals.

The implantation observed in identical twins can have a very different story. Some identical twins have been found at birth to be tucked inside a single placenta. Both are happily swimming together in the same fluid, two umbilical cords uniting into a single pipeline to the placenta. This fact tells us something about when the single fertilized egg initially split to form the twins. Since they shared a placenta, the preembryo was a single entity when it implanted into the uterus. The split had to occur after the raisin pudding cells tethered the preembryo to its uterine home. Thus, human preembryos are perfectly capable of splitting in half and creating viable humans at complex stages of development. This ability is rather reminiscent of the cow experiment mentioned in the last chapter. In that experiment, a blastocyst was deliberately cut in two, each half was placed into a surrogate, and identical twins were born as a result.

A shared placenta is not the only way identical twins have been found at birth, however. About 33 percent are born encased in two separate placentas. They are sequestered from each other, like they'd already gotten into a fight and had to stay in separate bedrooms. Since each possesses a placenta, and thus separate room service, exactly two preembryos embedded into the uterine wall instead of one. For that to be true, the split in the single fertilized egg had to occur before implantation. This fact illustrates that human preembryos are perfectly capable of splitting in half at very early stages of development and creating viable humans. It is also reminiscent of an animal experiment mentioned in the last chapter. In that experiment, a perfectly normal lamb was developed, using only the genetic information from a section of a very early stage embryo.

In the first ten days of life, the major preoccupation of a fertilized egg is growing and getting food. The first task of the preembryo is to divide; the first task of the mother is to escort the little passenger into the uterus. After it has reached the uterus, the developmental program begins hooking up the plumbing between fetus and mother. Exactly how many passengers will be drinking from the maternal trough is a question that can be up in the air for several weeks after conception. For the majority, a single preembryo will be created. But for some, multiple preembryos will be made. This fact has shed some intriguing light on the developmental capacities of fertilized eggs. These capabilities are rather reminiscent of animal experiments, where only portions of embryos were used to create entire organisms.

Human Relations, Part I

So, does a description of this biology get us any closer to deciding what is human life and what is human tissue? What we have so far is a ball of cells. The cells don't appear very much like a human being; they don't appear that different from a starfish. Arms and legs and opposable thumbs won't be coming for a number of weeks yet. A brain that can carry its own wavelengths won't be available for almost three more months. Based solely on appearance, distinguishing a human preembryo from that of a salamander, pig, or chicken is almost impossible.

The suspicion that this unfamiliar being is not a human is reinforced when we examine the fate of the preembryonic cells. In the beginning stages, the vast majority of cells will never be used to create a functioning baby. Even after considerable cell division, human life teeters on the whim of only a few cells. They will become a baby because of their physical placement inside the preembryo. If they were on the outside, they would probably become afterbirth, just like any outer cell. Thus, we see how the body distinguishes human tissue from human life: if a cell is on the outside, it will be a tissue; if a cell is on the inside, it will be a human.

Even if we decided that the appearance at ten days was sufficiently human, we still don't know how many humans we are describing. A fertilized egg could be a single human being; it could be many. The appearance of two or three or one is discernible at this point only by statistics. The ambiguities that multiple preembryos give us are the same ambiguities that single ones do. Physical evidence for the shape of the final human being appears to be in the future. Only the genetic imprint is present at such an early stage in development.

So what do we do? If we used physical appearance as our criterion, this preembryo would contain no more humanity than a sewing machine. The frustrating thing is that this preembryo is not always going to appear as a nondescript ball of cells. And it will never be a sewing machine. We are left with the disturbing possibility that an examination of the early events tells us nothing. For those who say that life begins at conception, the developmental clock is already set, and this discussion is pointless. But since certain kinds of preembryonic life do not have to begin at conception, it may be better to look at the end results of fertilization than at the beginning. The best question to ask may be, Can a fertilized egg grow up to be anything other than a single human being?

To understand the answer to that question, we must explore what the developmental program has next in store. The preembryo, now firmly anchored to its cellular harbor, is free to concentrate on refitting. It is only about ten days old. Which genes are going to be activated to turn these cells into muscles and nerves and bones can now be investigated. What kind of humanity exists at this time is one question. How we define its appearance may be an entirely different one.

Molecular Inner Tubing

The preembryo takes full responsibility for what it will call human tissue and what it will call human being. Given the mind-set of the twentieth century, it may be the last time it ever does so. It made that decision without first consulting the mother about what appeared human to her and what did not. That's a good thing because at this point the preembryo looks like a ball bearing. Now we must see how this nondescript roundness begins to paint itself into something with arms and thoughts and a beating heart. The postimplantation process is the subject of this section.

What does the body do with a ten-day-old passenger? Right now it is an embryo that has been embedded into the side of the uterus like a golf ball in a sand trap. This ball is composed of an outer group of cells, some of which have invaded the inside of the uterus. These outer cells form the placenta, an organ that ends up serving as a food delivery specialist between mom and baby. We also have that critical group of cells, the inner cell mass. This mass is stuck to one side of the ball's interior surface and will eventually form the fetus. The inner cell mass is only a few cells in population, much fewer in number than the outer cells. It will soon begin dividing like crazed rabbits, however, and as a result will split into two fairly defined layers. By the time the implantation has finished, its population is much greater than a few cells in size.

The other cells have grown in number, too. The outer cells that were busy making the placenta have become legion. These cells begin secreting the hormone called *hCG* about two weeks after conception. hCG is an example of science's inexplicable use of lowercase and uppercase letters; it stands for *human chorionic gonadotropin*. This hCG can be detected in the mother's urine very soon after implantation; its presence or absence is the basis for the results of most over-the-counter pregnancy tests. Another protein, called *EPF*, is made shortly after fertilization. It can be used as a marker to detect pregnancy in the first week of development. By the time the third week rolls around, the woman carrying the child may have skipped her first period, anyway. Detection of most pregnancies can occur before the end of the first month postconception.

The Layers of a Righteous Man

By the end of the second week, the shape of the inner cell mass has turned into something resembling a figure eight (see fig. 7.11). A yolk sac occupies the bottom circle of the figure eight. As has been stated, the embryo can grab a bite to eat from this sac but only for a while.

The top of the figure eight is reserved for a sac that will eventually cover the embryo like cling wrap covers yesterday's meat loaf. It is called

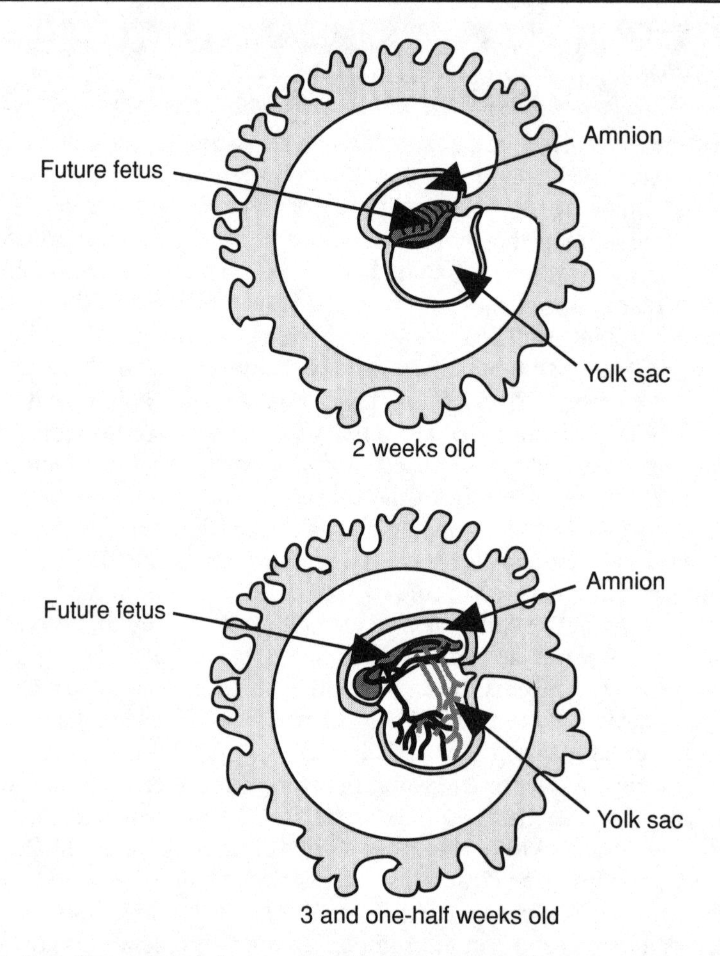

Fig. 7.11. **Human development in the first month. Note the presence of a figure eight-shaped group of cavities in the two-week-old embryo. The juncture of those two cavities possesses the cells that will eventually become a human fetus. The more familiar outlines of the embryo are seen at three and one-half weeks.**

the *amnion* right now and is technically not a sac. In the embryologist's never-ending struggle to confuse dentists, it is termed a *cavity.* This cavity will eventually get filled with a liquid, the *amniotic fluid.* This fluid can be used to examine the fetal chromosomes, which get dumped into it from the fetal cell. This technique is *amniocentesis* by which the genetic disposition of the fetus prior to birth can be judged.

So if the bottom of the figure eight is useless yolk and the top is cling wrap, where is the embryo? Or rather what is the embryo? At this stage, it is a thickened mass of cells that has broken into two layers. This bilayer exists right where the edges of the top and bottom circles of the figure eight touch each other. If you could somehow stand on top of the figure eight and look down, this inner cell mass would resemble a teardrop. The teardrop-shaped mass is the construction site where cells will be transformed into a fetus (see fig. 7.12).

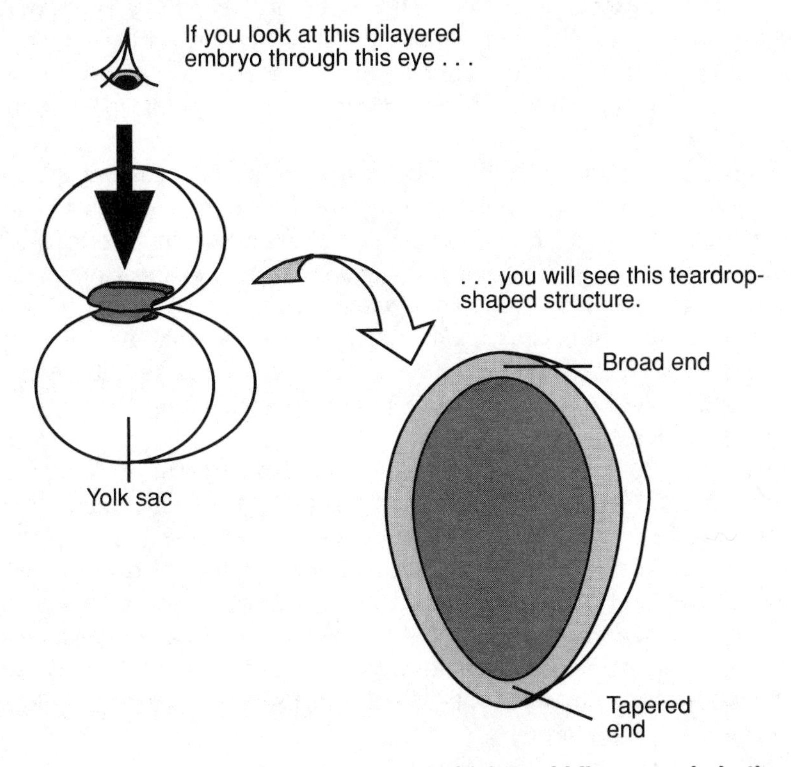

***Fig. 7.12.* The human embryo, seen as a thickened bilayer, early in the third week of life**

This section is devoted to understanding what happens to the inner cell mass in the third week of development, an extremely critical period when the bilayered embryo will turn into three layers. The foundations for every tissue in a human being will be laid down as a result. The embryo will begin the week with no defined organ system. It will end the week with a functional heart and a working circulatory system. To understand this critical part of human assembly, we really need to address the genes under the full sway of the developmental program.

And to do all that, we have to talk about homemade deli sandwiches.

On Being Well-Bread

I like a good deli sandwich, the kind stuffed with every agricultural product known to Western farming practices. I really like deli sandwiches that have been made with freshly baked bread. My wife is very good at making such bread. We have gone to the grocery store, bought an inordinate amount of stuffing ingredients, and then created sandwiches with her bread. Those sandwiches are second only to pizza and Enoch in terms of their brush with the divine.

I have tried to make freshly baked bread. It is a true culinary disaster. My bread is thick and gooey, and it often comes out in weird shapes. Sometimes my bread looks like a football suffering from skin cancer. Sometimes my bread looks like the vast wreckage of a landscape ravaged by a recently erupted volcano. For some reason, one frequent morphology that comes out of the oven is a teardrop-shaped loaf, reminiscent of the inner cell mass, broad at one end and slightly tapered at the other. These weird shapes have never prevented me from cutting the loaf in half and making a giant deli sandwich.

Now what do deli sandwiches have to do with the third week of human development? Nothing really. Nothing except for the shape of the bread I manage to make so poorly. As I mentioned before, this teardrop shape looks a lot like the embryo at the beginning of the third week with a tapered end and a broader-shaped end. There are two halves to this embryo, just like the sandwich that was cut in half. Unlike my sandwich, however, it is not filled with anything yet. The entire job of the third week is to make something that looks like a weird sandwich end up looking a little bit like a human being.

The Streak Shall Inherit the Earth

What happens next? We kick off the whole process by examining something that occurs on the surface of the embryo. For reasons that are not at all clear, certain cells from the edges of the embryo begin migrating to its lower center, like dancers lining up in the middle of a stage. Actually, it's more of a pile than a line, although it doesn't stay a pile long. After a period of time, this group of cells decide to thin out by migrating up and down the embryo, which causes a streak to be formed in the lower middle section of the preembryo. This streak is called the *primitive streak* (see fig. 7.13).

As more cells converge, the streak gets longer and longer. Soon a depression, called the *primitive groove,* is formed in the middle of the streak and extends more than halfway up the embryo. At the tip of the streak closest to the broad edge of the embryo, a little knob is observed.

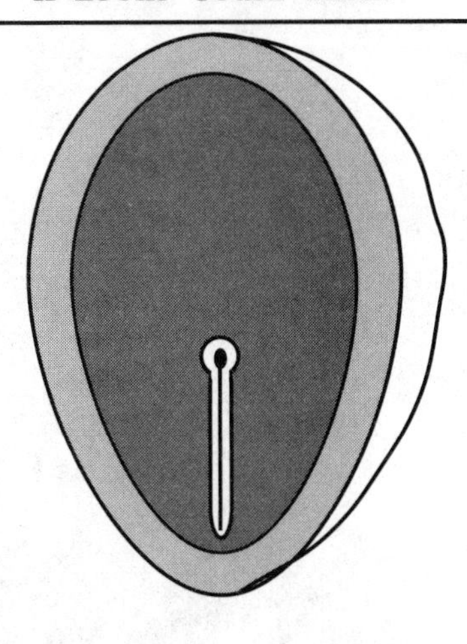

16-day-old embryo

Fig. 7.13. **Formation of the primitive streak. Groups of cells replicate and pile on top of one another near the tapered end of the embryo. These cells begin to migrate, forming a line of cells termed the *primitive streak.***

Under the microscope, the whole thing looks like someone had taken a finger and drawn a line down the lower back of the embryo.

Why does this streak form? The answer is not really known. It has been observed that as the groove forms in the center, cells from all over the embryo become attracted to it. Cells begin moving, like water molecules down a sink that has just been unplugged, toward the groove. When these migrating cells get to the upper lip of the groove, they splash down the ridge, like those same water molecules splashing down a cliff. The primitive groove is thus a continually changing cascade of cells, with wave after wave pouring into the ditch (see fig. 7.14).

Change Gangs

If this activity of the cells were to occur for an extended period of time, the ditch would fill up with lots of cells and no more cells would be left in the embryo. Neither is the case, fortunately. The embryo is growing, so there will be a plentiful supply of cells. And when the migrating cells splash down to the bottom of the primitive groove, they disappear into the interior of the embryo! Like the bottom of the ditch was made of quicksand or is a subduction zone, if you are into plate tectonics. This disappearing assures us that the ditch will never be filled. The only question is, What happens to the cells that get swallowed once they arrive at the bottom?

The fate of the swallowed cells is one of the most amazing processes in all of embryology. The cells, once they are in the subterranean atmo-

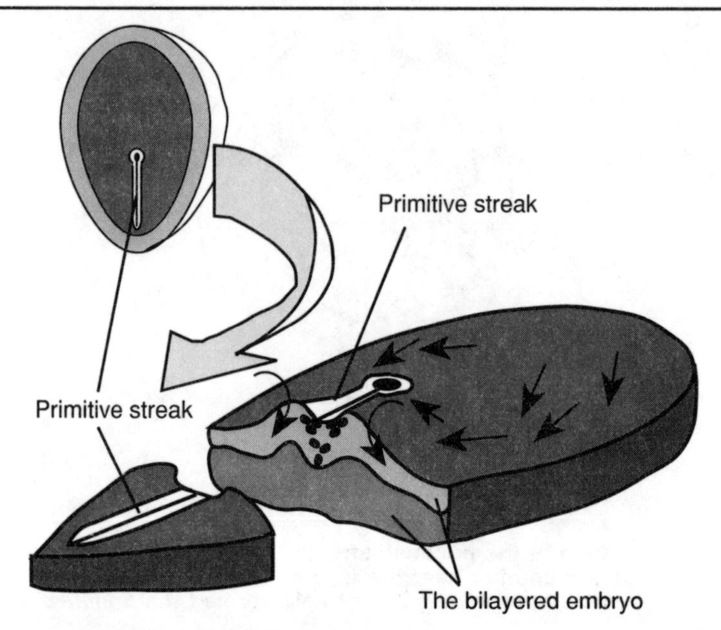

Fig. 7.14. **Migration of cells into the primitive streak. Cells from all parts of the embryo (shown by the black arrows) begin migrating into the primitive streak. Note that the orientation of the embryo has been tilted and cut away to expose the sites of cellular movement.**

sphere of the interior, begin to change their appearance, like caterpillars changing into butterflies.

What these cells change into depends on what part of the ditch they originally fell through. One can roughly divide the ditch into two regions: the region closest to the tapered end of the teardrop-shaped embryo and the region closest to the broad-edge (the end that has the little knob). We will examine the fates of cells swallowed in these regions one at a time.

A Ditch in Time, Part I

Let's first examine the fate of the cells that fell through the ditch near the tapered edge. Once inside, the cells continue to move and to divide; they spread out and grow in the interior of the preembryo. But because the preembryo comprises two layers of cells, just like a sandwich has two pieces of bread, the cells that were swallowed are not allowed to penetrate the lower layer. As the cells continue to move and divide, they fill up the middle space between the top and bottom layers, and the middle layer gets thicker and thicker. Kind of like a deli sandwich that gets thicker and thicker as you put more meats and cheeses and vegetables in its middle.

This new middle filling is extremely important when examined with the original top and bottom parts of the embryo. Three specific layers of cells are established: the top layer, *ectoderm*; the bottom layer, *endoderm*; and the new filling, *mesoderm* (see fig. 7.15). (*Derm* means "skin," which is a lie in that these cells are no more skin as we think of it than is a Toyota pickup.) Specific organs and tissues will arise from each layer (more on that later), and an entire baby will be formed.

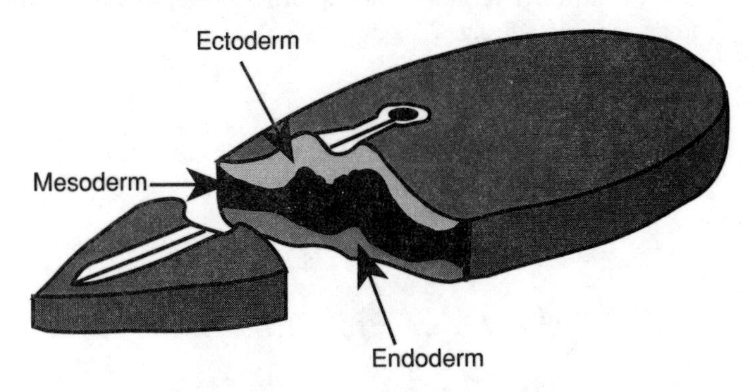

Fig. 7.15. **Formation of the three embryonic layers. Each layer of cells will give rise to different organs and tissues in the developing embryo.**

A Ditch in Time, Part II

What about the cells that enter the ditch closest to the broad edge of the preembryo, the part that contains that little knob? On close inspection, we see that it is not a knob at all; rather, it is a depression or well. As cells migrate down this well, the well doesn't fill up. Instead it gets deeper and deeper as the migrating cells form the sides of a self-burrowing tunnel that goes straight down through the first layer of the embryo.

But these cells cannot burrow through the embryo and create a gaping hole. They are stopped, like their cousins, between the top and bottom layers of the embryonic bread. The second layer proves to be as much a barrier for the tunneling cells as it does for the cells at the tapered end. But only the direction of their tunneling is deterred. The cells are as determined to make a tunnel as Congress is determined to make a deficit. Rather than give up, these cells simply turn 90° and continue creating an underground mine shaft. The shaft moves toward the broad edge of the embryo, forming a tube between the two halves of the embryonic layers of cells. As the cells continue to reproduce and penetrate the interior, they form an internal tube, similar to a central underground pipe, up the center of the embryo. This pipe is sandwiched directly between the two halves of the embryo.

There is an end to all this cellular digging, even for such determined cells. The tunneling cells eventually encounter a barrier near the edge of the broad end of the embryo. And this time the cells do not turn 90° or go around the barrier. They lose their desire to burrow, and as a result, they halt their tunneling. The barrier that stopped this tunnel from going forward is the *prochordal plate,* which is part of a group of cells that will eventually become the oral cavity of the developing baby. This is an example, hardly unique, of a biological exploration that has been stopped by somebody's mouth (see fig. 7.16).

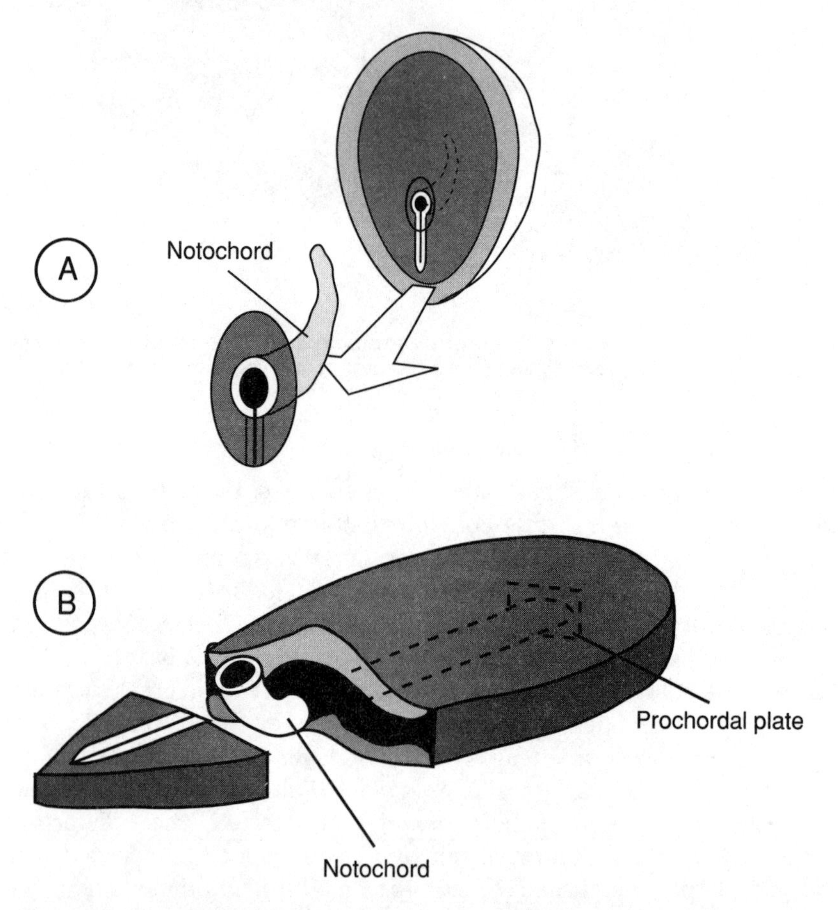

Fig. 7.16. **Two views of the relative positions of embryonic tissues and notochord. The teardrop-shaped embryo (A) has been rotated and cut away to expose underlying cells (B). Notochord is shown, relative to surrounding tissues, larger than actual.**

Self-Policing Genetics

So who are the traffic cops that coordinate all this cellular movement? External molecules and forces, as well as the internal biochemistries of the individual cells, supervise the activities. For example, a cell moves because certain internal genes create proteins that can respond to external orders to move and can physically get up and go. A cell disappears inside the bottom of the primitive streak because the genes inside told it to, probably in response to a variety of external cues. One cell layer halts the movement of an individual cell because the halter was programmed to stop the cell and the haltee was programmed to respond. Most of these moving, disappearing, and halting genes were activated by substances produced by genes in still other cells. The other cells were either right next to their target or far away.

Remember the alarm clock analogy? Certain genes become awakened in certain cells, which then awaken still other genes in other cells in a predetermined—often precise—order. This wake-up call gives them the ability to move, disappear, or halt. We are only beginning to understand the nature of these calls and signals. Only certain cells will respond to these summonses at any one time. If all the cells made ditches, for example, we might be born looking more like balls of wound-up string than human beings. The whole thing is an incredible, mind-boggling tapestry of interconnecting biochemistries. At this stage in our technology, most of it is also a complete mystery.

The Chord of Directors

So far we have an underground tunnel that starts around the middle of the embryo and finishes underneath it, nearly at its far edge. For you vocabulary buffs, this tunnel is called the *notochord*. The space in between the two halves of the embryo is beginning to get quite crowded. The middle layer is forming from the ditch cells near the tapered end. Now an underground canal, this notochord, has been established between the two cell halves. The spreading cells that provide the middle filling (from the tapered end) don't seem to mind its presence, however. They swarm around the canal even as it forms.

You may be wondering why any of this underground tunneling occurs. Most of us wonder at it, too. Much of its internal biochemistry is a complete mystery, a fact made worse by demonstrating that the notochord can be removed from a lot of animals and an embryo will form anyway. A hint may come from the close examination of the cells at the surface of the embryo immediately above the tunnel. As the tunnel forms beneath them, the cells at the top biochemically feel all the commotion and, in so doing, react to it. How might they react? By forming another canal, this time on the surface. A raised groove is constructed directly above the underground tunnel on the surface of the embryo. These surface cells may use the notochord as a guide to create this surface canal. It's a bit like a mole; you can see where a mole has been tunneling by examining the reaction of the surface immediately above its tunnels.

Whatever the mechanism, the genetic program eventually activates the cells in this new canal to become the central nervous system. The head end of the canal will become the brain. The rest of the canal on top of the subterranean tunnel will become the spinal cord. There is a problem with this arrangement, though. To place these organs in the body correctly, these cells also will have to move under the surface of the embryo. You would not be very functional if your spinal cord lay directly on top of the skin of your back. Realizing this, the canal begins to sink. As it sinks into the embryo, the canal curls on itself. By the time it completely submerges, the groove closes and is a tube (see fig. 7.17).

How far does it sink? It stops just above the tunnel, the notochord, that guided its architecture in the first place. Washing over it on top, like the waves of an ocean, are more surface cells. This ocean, now lying on top of the future spinal cord, will eventually become the skin of the embryo.

Falling on Heart Times

By investigating these overall processes, we can see that much of the foundation of the embryo is laid in the third week of development. The three layers that will form most of the human tissues, from organs to integument, are all in place. The spinal column and the brain have gotten a head start. The embryo is beginning to appear less like a nondescript teardrop and more like a living organism.

We haven't examined how the embryonic cells get nourishment because we haven't yet touched on the manufacture of the heart and attendant blood vessels. So, next let's discuss the formation of the *vascular tree*. We will start not by peering into the inner universe of cardiac development but by looking at the outer universe of stars and constellations.

When I was little, I saw a movie in school that taught us how to wrest the constellations from those unorganized groups of stars in the night sky. Ever since then I have enjoyed transforming the night sky into high drama, seeing not distant orbs of burning gas but arrows and dragons and kings and queens. Moreover, I am in constant awe that we are not looking at things as they are; we are beholding all at once many chapters from a cosmic history book. Those thoughts always make me shiver. As a result, I usually am content to find the familiar Big Dipper and Orion and leave the cosmic questions to better minds.

Many analogies can be derived from this ordering of stars to the developing embryo. The magic of biological development is that the cellular

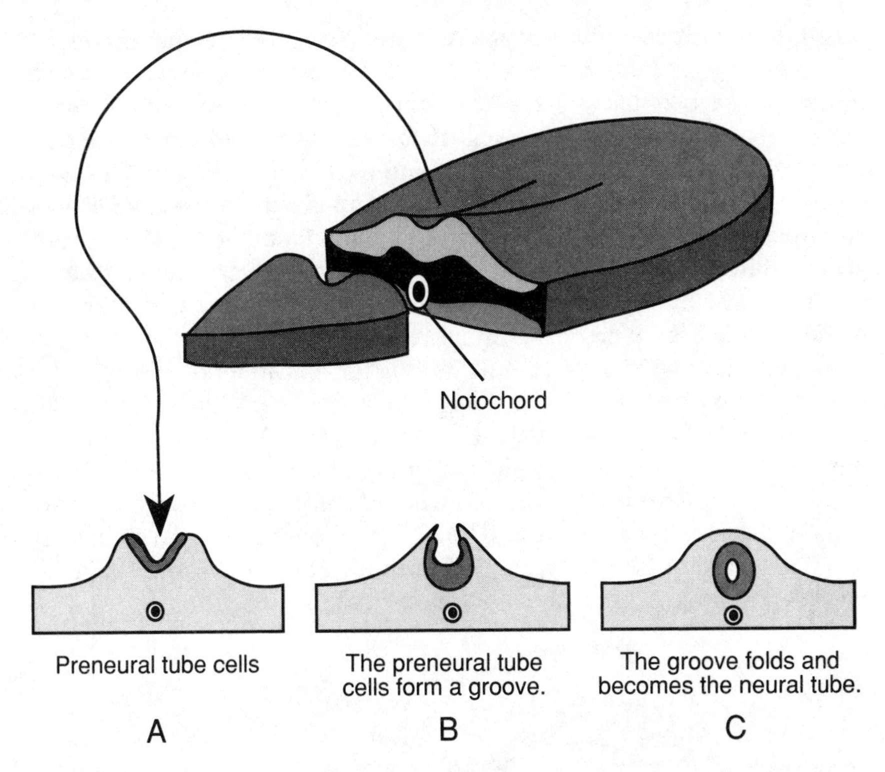

***Fig. 7.17.* Formation of neural tube. A cross section of a progressively older embryo, ending at four weeks, is shown in panels A through C.**

architecture does not have to exist only in our imaginations. If we wait long enough, the distant cells will push themselves into a recognizable form, like a living constellation, without mental assistance. One can follow the history of human development by looking at early pictures of embryos and comparing them to later ones. Such a developmental process is witnessed in the makings of the circulatory system.

Willing Vessels

Detecting a circulatory system at the beginning of the third week is not impossible. In evidence are individual and solitary cells that have their origins in one of those surface grooves discussed earlier. These primitive cells migrate to many parts of the embryo, like single stars flung across an embryonic sky. As more of them are made, they become sticky and adhere to each other.

In a fairly short time, many clusters of cells are scattered throughout the embryo. These clusters are termed *blood islands*. The cells in these islands begin to divide and form little rings of cells. Under the microscope, they look like tiny donuts. As more sticky cells come in contact with these donuts, the donuts elongate and look more like tubes. These tubes find each other and fuse to form a complex series of channels. These channels are the beginning of blood vessels, the veins, arteries, and capillaries that will feed the fetus and eventually the adult. This connecting of clusters is a little bit like drawing lines between stars to form a constellation. By the end of the third week, the blood vessels have been arranged in sufficient complexity to form a primitive circulatory system.

While these blood islands are forming vessels, two very special sets of blood islands are made in the broad region of the teardrop-shaped embryo, and they will form the heart. Starting out as two tubes lying side by side, the tubes fuse and form a single tubularlike heart. By the end of the third week, the cells have begun beating, a task they will continue to perform for almost seventy years. Other blood vessels now fuse with the new heart. Even blood cells are synthesized (probably from those primitive donuts) and migrate into the heart and primitive vascular system. A functional cardiovascular system is thus made. It is the first system in the entire embryonic body that is operational (see fig. 7.18).

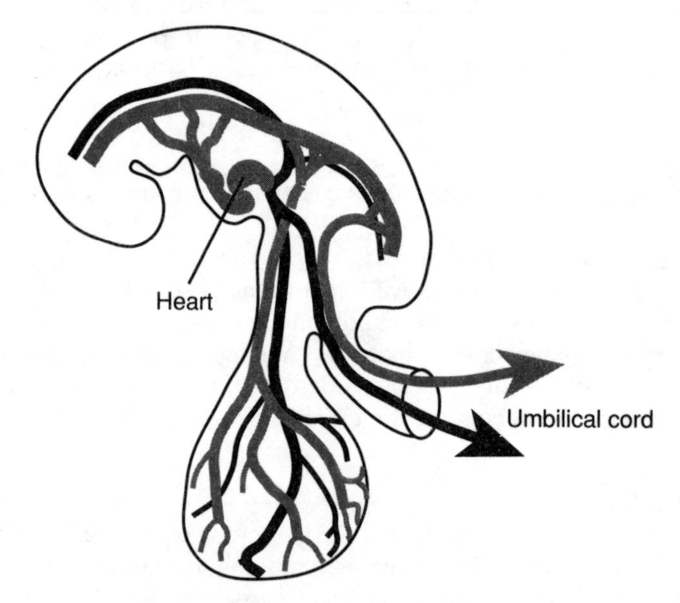

Fig. 7.18. **Cardiovascular system (about three and one-half weeks)**

Human Relations, Part II

By the end of the third week, we have all the tissues needed to form a baby and a way to feed them as it matures. This organization then begs a familiar question. Does a description of this biology necessarily get us any closer to deciding what is human life and what is human tissue? The embryo is a three-layered teardrop with a line up its middle and a heart that beats. This creature doesn't look very much like a human being. At this stage the embryo looks more like a segmented worm. The fate of individual cells can be questioned because most tissues of this organism are still placenta.

If we used only physical appearance as our criterion, we would say that this embryo contains no more humanity than the worm. However, the embryo has begun to take on characteristics that will processively separate its physical appearance from any other animal, including night crawlers. Shall we define protectable humanity when it becomes more physically familiar to us? If so, what are the physical criteria we will use? A spinal cord is familiar to us. So is a heart. Such organs are also familiar to birds and camels. At the organ level, nothing is yet uniquely human about a three-week-old embryo to separate it from birds and camels.

We are left with the disturbing possibility that this stage may tell us nothing about protectable humanity. Since there is so much that looks inhuman and will never become fetal, it may be better to look at later stages of development. Can this three-week-old embryo grow up to be anything other than a human being?

To understand the answer to that question, we must explore the next brush strokes of the developmental program. We have already come a long way from a single fertilized egg. In one month, the embryo has presented all the cellular colors necessary to create the basic tissues of the human body. The developmental program is now ready to enlarge those strokes into a full-fledged portrait.

Dilutions of Grandeur

I knew a well-traveled man named Tom. He was a very bright and cultured person who, in his later years, began to suffer from a mental illness. He slowly came under the delusion that he was a great painter of portraits. I know a number of artists who feel that way about their work. They suffer from the same delusion, although they are not mentally ill. But my friend Tom believed he was one of the masters. He would describe the Matisses he'd seen, the Renoirs, the Rembrandts, all as his own work. He would invite people to go to the various museums—he knew them all—to view his artistic accomplishments. When you asked him if he had ever lifted a brush in his life, his quick reply was no. Tom had never painted a portrait in his life. But that did not stop him from claiming his territory as a great artist because he had visited the museums that housed the paintings.

I always think of Tom when I think of creativity and personal responsibility. His portrait delusion invades my thoughts when I think of embryology, too. In a way, creating a baby is a little bit like drawing a biochemical portrait of oneself. Only there are two artists. And the portrait is alive and fully interactive with the artists who created it. Unlike painting, parenting is an art whose entire goal is wiping out its reason for existence. The task is to make this living portrait increasingly independent of parental input and need. There is even a time when the parents must let go of the child, a task I'm told is one of the hardest things to do in child rearing. When it's all said and done, you realize that you have been in the great museum of parenthood ultimately as a viewer, rather than a creator, of the work. For some parents, this realization is so abhorrent that they reject it. Like my friend Tom, such parents take responsibility for a portrait that was never really their own.

Mothers seem to have that curious mixture of responsibility and remoteness during their pregnancy. No wonder. Even by twenty-one days, most of the specific cellular material necessary to create an entire human is in place in the embryo. Not bad for a few weeks' work. We can talk about the rest of the development story in more general terms. The purpose of this section is to briefly summarize what happens between the second month and the ninth month postconception. And perhaps to shed some light on both the remoteness and the responsibility we often take in the investigation of our own kind.

Cellular Phones

Much of the process of human development is really a story about communication between molecules. The local and long-distance calls genes make to various cells in the embryo determine how the embryo will turn out. The fates of individual cells in the three layers of cells hinge on this communication. There's even a time constraint. Within nine months, a fully functional human being must be formed and in operational readiness. To me, this communication circumscribes a double miracle. The first miracle is that it happens. The second miracle is that once it starts, it proceeds correctly and more or less on schedule.

Communication occurs between genes and the cells housing them. As we have learned, genes need to be awakened and put to use to function. In the beginning stages of life, certain active genes exert vast powers over the developmental future of the embryo because all of a human being is locked inside a few cells. The master switches, when they are stimulated, affect everything the developing embryo will become.

As development proceeds, increasingly specialized cells are made.

Genes become activated that basically give career counseling lectures to these cells, forcing them to commit to a certain line of work. Such genes limit the options that a particular cell can choose when it grows up. For example, some genes control the beating of heart cells. Cells that become committed to a cardiac future have beating genes awakened; these cells lose their ability to become brain cells or bone cells or what have you. The process of cellular development is an exercise in an increasing loss of occupational flexibility. Eventually, everyone becomes specialized as the direct result of the communication of genes to the cells that bear them.

Making Lenses Meet

Another form of communication exists between cells in specific locations. The genes in some cells can produce products to tell other cells what to do. For example, the cells that form the very beginning of the eye are collectively termed the *optic vesicle*. If an optic vesicle is present in the top layer of our deli-sandwich embryo, it will induce another structure of the eye, namely, the lens, to form right next to it. Why? Because molecules produced by genes in the optic vesicle escape it and go on to communicate with the cells of the embryo's top layer. If the optic vesicle is removed, the lens will never form in the top layer because its cells will never receive those molecules. In this example, the communication is one-way; the optic vesicle gives instructions, and the top layer receives them.

But let's say this vesicle is taken from the top layer and put somewhere else, say, where a foot is supposed to develop. The genes in the optic vesicle will induce a lens to grow in the foot! The optic molecules communicate a potent message to neighboring cells, regardless of where they might be. This incredible plasticity is an example of *inductive interaction*. Such cell-cell interactions are an extremely important part of the development of the human fetus, and indeed, developing cells in general.

This foot-lens experiment highlights another principle about embryonic growth mechanisms: development is a combination of specific genetic instructions and the positional information of cells. We don't normally make eyes in the middle of our feet. Thus, the genes making that powerful optic vesicle are never turned on near our ankle bones. But they could be. Ankle bone cells contain all the genetic information to make a complete eye. The task of the developmental program is to activate specific genes only at certain sites. Otherwise, we will be born looking less like humans and more like Picasso paintings. The embryo has to sense the position of each of its cells, up from down, left from right, backward from forward. It must then use that information to incite specific genetic signals at specific places. And only at certain times. As more cells come into existence, the developmental program must shepherd greater and greater numbers of processes. The whole thing is unimaginably complex.

This specialization is never more obvious than in the three layers made during the third week. You recall that a group of cells invaded the in-between space of the top and bottom halves of our embryonic loaf. This resulted in the formation of those three layers, the top, the bottom, and the middle. Each layer makes more or less specific cell types, although the fates of each are not rigidly fixed. Let's see what happens to each layer.

The Top Layer

The top layer of cells in our sandwich (the ectoderm) is responsible for making a variety of cells, mostly having to do with nerve tissue. You recall that this top layer may respond to a tunnel dug underneath it like the surface of the earth responds to a mole digging underground. In this case, the top layer makes a canal that will eventually sink deep into the embryo. This canal will form the spinal column and brain. The cells that wash over the top of this sunken tube will create parts of the skin. This top layer also is responsible for the creation of other cell types. Briefly summarized, the top layer will give rise to the following tissues: (1) brain; (2) spinal column; (3) skin, hair, nails, breasts, and sweat glands; (4) inner ear; (5) eye (retina and lens); and (6) enamel of teeth.

The Middle Layer

The middle layer of cells in our sandwich (the mesoderm) is responsible for making the muscles, skeleton, and various internal organs. You recall that this middle layer was created when some of the cells near the tapered end of the embryo decided to plunge into its interior. These cells soon filled the inside of the two-layered embryo like meats and cheeses fill the inside of a deli sandwich. Briefly summarized, the middle layer will give rise to the following tissues: (1) muscles of trunk; (2) skeleton; (3) gonads; (4) blood and lymph; (5) spleen; and (6) heart.

The Bottom Layer

The last group of cells is that bottom layer in our sandwich, the endoderm. You recall that this bottom layer was resistant to penetration from certain cells attempting to invade it from above. This stubborn group of cells has the responsibility of creating a specialized group of tissues, the *epithelial lining,* which exists in most organ systems. Epithelial linings are usually single layers of closely packed cells on the internal/external surface of organs. Such cells resemble the inside lining of a trench coat and have a variety of critical functions. Briefly summarized, the bottom layer will give rise to the lining of the following tissues: (1) lungs; (2) throat; (3) liver; (4) genitourinary tract; (5) pancreas; and (6) urinary bladder.

These tissues follow their own course of development. Different organ systems come up at different times as the embryo matures into a fetus. They all have to work together to form a functional organism. We have already seen that a workable heart is created before the first month has closed. To keep track of such development, we have arbitrarily divided human development into three chronological stages called *trimesters*, each three months or so in length.

The Foundation Trilogy

The first ninety days of growth are characterized by profound changes in the shape, number, and composition of cells of the embryo. At the beginning of the first trimester, the embryo is exactly one cell in size, barely the stature of a speck of dust. By the end of the first trimester, the fetus possesses a miniature human shape almost a quarter inch in length. As we have seen, the first thirty days of this growth establish the developmental foundations for the rest of the pregnancy. All ninety days are important, however. Most severe deformities in development happen during these first three months.

The most dramatic change in the first trimester is the shape of the embryo. It goes from a nearly microscopic basketball to a highly visible C-shape, complete with little arms and legs. The embryo has to do a lot of stretching and bending in three months, contortions necessary if the fetus is to assume its familiar profile. For example, we start out with the beginnings of the head stuck onto the neck, much like a fish. The primitive eyes are on either side of the primitive head. The body is a straight rod. If the face growing on the top of this rod never changed its angle, it would continue to develop on the top of the head. The developmental program issues instructions to bend the neck in the first trimester. Any tissue growing out of its head will be 90° to the vertical plane of the spinal cord, including the location of the eyes.

The eyes have another positional chore to perform at this stage, however. The newborn ends up with the eyes at the front of the face directly above the nose rather than at the sides next to the ears. Since they start out on the sides of the head, the eyes have to "migrate" to the front of the face. By the beginning of the tenth week, the eyes are positioned squarely on the face.

Shape is not the only thing that occupies the fetus's time. The genes that govern the development of the internal organs are running full tilt in the first trimester. The heart has started beating. In a month, a fully functional circulatory system is observed. The internal workings of the eyes form during this time. The genes that will guide the construction of the outer ear have begun instructing cells to pile up on each other, one pile on either side of the head. By the end of the eighth week, they have formed a definite and observable ridge.

The developmental program at this time dips into its stock of middle-layered cells and instructs them to begin creating muscles. The intestines take shape. Urine forms, a fluid swallowed by the fetus (with no apparent harm) or shuttled out of the placenta and given to mom. The brain has begun to develop to such a point that by the twelfth week, a brain wave separate from the mother can be detected in the fetus. Clearly, first-trimester growth involves changes in external appearance and internal anatomy.

A few cryptic structures form in the first-trimester embryo. As the skeleton forms, a real, live, completely visible tail is seen (usually by the fourth week). This tail eventually shrinks and by the seventh week has completely disappeared. Why is it there? Except to intensify the fight between young earth creationists and staunch Darwinian evolutionists, nobody really knows. We also get what appears to be gill arches near the mouth, structures normally seen only in fish, by the fourth week. During a period of time in the middle of this development, the fetus looks more like a tadpole than a human. Those gill slits eventually disappear, and by the end of the first trimester, the fetus has a distinctly human appearance.

The developmental program has done enough genetic sculpting that by the second trimester, various behavior patterns have begun to be formed. For example, if the lips of a twelve-week-old fetus are stroked, the fetus responds by sucking. Gently touching the eyelids of that same fetus will cause a reflex reaction to occur. These responses demonstrate the extent of nerve/muscle wiring already in place by the beginning of the fourth month.

Biological Growth Industries

These reflexes usher in the second trimester, a period characterized primarily by a tremendous burst of fetal growth. If the pregnant mother grew at the same rate as her baby at this stage, she would be as tall as a two-story house and weigh more than a ton. Such immense growth means that when the baby decides to move, the mother will feel it. These movements include gentle rhythmic kicks, not-so-gentle nonrhythmic kicks, and lots of squirming.

This growth is also accompanied by the further development of tissues. The inner ear will be fully developed in the second trimester. As it is hooked up to the brain, it becomes remarkably responsive to sounds. For example, the *in utero* fetus can react to loud noises such as the slamming of a door and the honking of a horn. It can react to sounds that cannot be heard by the adult human ear, suggesting the presence of a sensory pathway not involving the ear. Scalp hair patterning becomes fixed at this time, giving a clue to the extent of brain development. As the tiny hands develop, the genes that make fingernails become activated, and fingernails begin to grow. Lung tissue is present, although it is immature and generally incapable of supporting sustained breathing. Babies who are prematurely born in the second trimester usually die because they cannot breathe.

The second-trimester fetus grows a number of tissues that go away prior to birth. For example, the fetus temporarily secretes a greasy, cheesy substance all over its body. This stuff is the equivalent of Chap Stick lip balm and protects the delicate fetal skin from abrasion and hardening during its stay in the uterus. In addition, it grows fine downy hair all over its body, which probably holds this grease onto the surface of the skin. The fetus also develops a tissue that is the equivalent of a temporary space heater. Known as *brown fat,* this stuff contains a tremendous store of energy in the form of mitochondria. You might recall that mitochondria was the same energy store sperm cells used to propel themselves into the uterus during conception. The brown fat usually disappears prior to the onset of the third trimester.

The main task of the fetus in the second trimester is to hook up tissues laid down in the first trimester and then get bigger. The developmental program, as usual, is busily activating genes. As the fetus grows, the program must additionally ensure that the organs are in the right place at the right time. Eventually, the presence and the integration of various complex tissues and organ systems are observed. There can be no misjudgment about what this organism is going to be. By the end of the second trimester, the overall physical appearance of the fetus is even more human.

Finishing School

The last three months of life in the womb are not devoted to major changes in the shape and chemistry of the fetus. There's little to initiate at this point that hasn't already been started and, in many cases, finished. The fetus fine-tunes and tests the genetic processes established in the previous six months. For example, an important task is learning to use its lungs. Even though the fetus is still bathed in fluid, the developmental program instructs that little diaphragm to begin practicing its expansions and contractions. Such movement is interrupted by sighs and, curiously, in utero hiccups. The eyes, also bathed in fluid, become fully open. They can blink. The fetus responds to the presence of bright lights by increasing its general activity. A calming effect on general fetal movement occurs when the womb is subjected to repeated flashings of light.

There are still some genetic changes, of course. Different tissues are built up to prepare the infant for life outside its uterine swimming pool. The manufacture of blood, which had been done primarily by the spleen, is turned over to the tissues that will do most of the job as an adult, namely, the bone marrow. In males, the testes are instructed to parachute into the scrotum from their former intestinal perch. The developmental program instructs the fetus to stop making brown fat and begin making normal white fat, stuff it will use and manufacture once outside. The fetus responds to this instruction with enthusiasm in its last month. The fetus will make almost a quarter pound of the greasy white stuff each week. By the time it is born, fully 16 percent of its body weight will be solid fat.

When the baby is fully insulated, when the lungs become like working bellows, the developmental program begins to loosen its grip on the fetus. There is a time when the infant will be forced, literally kicking and screaming, to leave. If the fetus stays too long, it can lose weight and develop an almost papyruslike skin. Such a condition, called *postmaturity syndrome,* is very unhealthy.

With the onset of birth, the finishing touches on this living portrait are completed. If the developmental program did its job correctly, it will give this baby the ability to live its threescore and ten. It will also confer the ability to ignite an entirely new developmental program with a member of the opposite sex. With these abilities firmly under genetic wraps, there is a sophisticated functionality associated with a human appearance by the end of the third trimester. This fully developed baby is complete with hands and toes and eyes and a brain.

Possession with Intent to Distribute

Getting back to a familiar endpoint is comfortable. We start with a single nonhuman being sperm and egg, and we end up with a single very human baby. We've been looking, however, for a developmental event somewhere in the middle that might give us a guidepost as to what is human being and what is human tissue. We have been examining physical appearances at various stages to see if any idea scurries out from under the embryonic rock and becomes conspicuous. The scary thing about this issue is that any conclusion we reach has the capability to cause a riot in the human heart. Or even in city hall or the nation's capital. Asking the question has the ability to make us feel extremely uncomfortable.

Why is that? After much experience in the realm of hypothesis testing, I have come to the conclusion that ideas are sometimes like children. It can be very difficult to give them up once we give birth to them. We tend to nurture them and resist suspicions, let alone investigations, of their validity. That is why technology, especially genetic technology, can be so controversial. The discovery of a new idea often means that an old idea is wrong or at least misdirected.

We didn't invent the pieces of data that might prove us wrong. We didn't invent the biology from which the data were extracted. Science gets its muscles because it investigates what already exists. The problem is in us. When we cannot let go of an idea that flies in the face of certain kinds of evidence, we suffer a delusion. It reminds me of my old friend Tom, who could not give up the notion that he was a great artist, even though he had never held a paintbrush.

Putting In an Appearance

If appearance was an adequate criterion to judge when tissue got its humanity, the organism would no longer be a human being when it no longer looked like a human being. The question would then become, What is the criterion for judging what a human being looks like?

We know that humans can take many forms. Dwarfs don't look like tall African Masai people. Germans don't normally look like Koreans. Your brother doesn't look like your sister. And I don't look like you. Even when the genetics are normally expressed, there are vast differences in our overall shapes and characteristics. Such differences are also seen in individuals whose genetics are abnormally expressed. Or in those who have undergone severe trauma. People with twisted, mangled arms and legs due to certain diseases do not share the twentieth-century appearance of most of us. There are burn patients who, though they survive their injuries, come into the emergency room barely recognizable as human beings. People who have had both arms and legs amputated look like a chest with only a head attached. Are these people less than human beings because they appear different from us?

Physical dissimilarities can be observed beyond external appearances. You and I share many common traits, mostly centering on our body plans and internal organs. But there are tremendous differences in the way our bodies are constructed at the biochemical level. The physical appearance of our genetic machinery is different. You cannot accept a little bit of my blood into your body without rigorous clinical testing. You cannot have my heart grafted into your chest without living for a lifetime in the presence of very powerful anti-rejection drugs. That's true even if the heart came from your mother or your uncle. Thus, dissimilarity due to appearance can be seen even at the level of cells and molecules. Are some of us less human than others because we do not biochemically look alike?

Human Relations, Part III

These issues must be taken into consideration when one attempts to answer the tissue/being question based solely on physical appearance. Perhaps the most powerful example of this distinction lies within the developing embryo. The organism may possess humanity without looking the part or may not possess humanity while fully looking the part. A single fertilized egg does not look like that beach ball-like blastocyst. A thirty-day-old embryo just barely resembles a newborn baby. A baby at nine months is almost five hundred times heavier than one at nine weeks. Are the organisms at any of these stages less human simply because they do not look like one another? Or more human simply because they do? Is humanity something that sort of creeps up gradually and then, at some fixed point, pounces on a group of cells? If physical appearance is part of the equation that solves our sentience question, we must determine exactly where it gets dialed in.

This determination is what's wrong with the idea of appearances. Deciding where protectable humanity commences based on looks is quite arbitrary. It is like attempting to determine the quality of a marriage by examining a silhouette of the husband and wife. We have spent this chapter examining the biological processes that govern the events of development. We do not find staccato, punctuated reactions that proceed in rigidly defined, compartmentalized steps. Instead, we find a beautiful unfolding of complex processes under the guidance of an extremely efficient genetic bureaucracy. These steps are gradual, with human form processively focusing into a recognizable shape as the pregnancy advances. If an attempt is made to paint a defined portrait of protectable humanity at a certain step, it must be hung on a continuum of developmental biology.

The contrary argument is also true. We can be just as tempted to call something human the moment we recognize a familiar shape. However, just because something looks like a human being does not mean that it is a human being. A corpse looks like a human being. So does a mannequin. Neither is sentient, protectable humanity. The same rule must be applied to the developing fetus. Just because it happens to look like a human does not mean that it is one. Or is going to be one.

In the next chapter, we will discuss certain deformities that boldly form a complete human shape and yet have no more sentience than a bowl of Jell-O. This will demonstrate what all of development illustrates: pronouncing value judgment on tissues by examining their biological appearance makes as much sense as judging how a candy bar tastes by eating only its wrapper.

We need another set of criteria. Or perhaps a better question. Examining specific developmental stages does not give us the answer. Is there a trait that defines unique human sentience that is not so dependent on appearance? We have repeatedly asked throughout this chapter if a developing embryo could be anything other than a human being. The answer to that question is, surprisingly, yes. To explain this answer, we have to explore a trait that some say forms the criteria for discerning sentience in cells. To do that, we have to describe the birth of a very special baby, a baby born normal in every way except one.

It is born without a mind.

Chapter 8

All Thought and Bothered

Introduction

The famous philosopher Descartes once said, "I think, therefore I am." That statement gives our heads a lot of credit. If that's where all human awareness exists, our educational system should be the Fort Knox of sentience. However, in the days when 40 percent of all college-bound high-school seniors cannot find themselves in the mirror, placing our actuality on the contents of our skulls is risky business. If it is true, Descartes's conclusion is quite alarming.

But perhaps we needn't be worried. A great contribution of neural science has been to show that Descartes was full of baloney. The problem was that he couldn't define thinking accurately. It wasn't his fault; he lived in a day when people were responsible for their actions and couldn't explain problems away by saying they needed more algae in their diets. On the other hand, he also lived in a day when the proper cure for epilepsy was drilling a hole into the cranium and pouring salt into the perforation.

If Descartes knew about the advances in modern twentieth-century neuroscience, he probably wouldn't have bothered with his definition anyway. Thought processes can be divided into so many parts that, as a concept, human thinking loses most of its original meaning. Thus, putting the entire weight of existence on an idea whose validity is suspect is as stupid as waging peace by making thermonuclear weapons.

Descartes's statement implies that the awareness of thinking, itself a thought, is the evidence of existence. To be fair, this idea still has something to say to us even after it has been scrubbed with twentieth-century medicine. Our ability to think in specific ways distinguishes us from every other biological organism on many levels—so many levels that it is a candidate for wresting human sentience from human tissue. Our task, in an attempt to do the same, is to try to define the contents of human thinking more accurately.

To accomplish this goal, we will first explore a birth defect called *anencephaly.* This deformity affects cerebral function and serves as a starting point to describe the contents of human thought. We will then attempt to summarize certain advances in neurology regarding thought processes. Beginning with the activity of a single nerve, we will study the biochemistry of memory and consciousness. With that in mind, we can return to Descartes and see if he deserved the drubbing neurochemistry has given him.

And to do all that, I would like to start from a not-so-obvious neurological point. I would like to begin with the biology of the ability to see.

The Tie that Blinds

I have always wondered what people who have been blind since birth do with words like *pastel.* Or *golden.* Or *sky blue.* This has always been intriguing to me because in many cases, the part of the brain used for seeing starts out intact in the visually impaired. It's usually the eyes, those marvelous shepherds of light, or their connections to the brain that are unable to bring anything coherent to their cerebral masters. Blind people, of course, form images and textures in their minds just like sighted people do. And they can have definite impressions and feelings about objects, and even colors, that they have never visually beheld. It brings up the interesting fact that human beings do not see with their eyes. Eyes only collect light. Human beings see with their brains.

This fact is illustrated very well by considering the following bit of biology. Let's say that you have just watched a college freshman swallow a goldfish. The light bouncing off the obnoxious young man and his hapless aquatic victim enters your eyes. The image immediately falls onto a collection of nerves at the back of your eye called the *retina.* Those cells, once excited by light, return the stimulatory compliment by sending off a signal to the brain. Because of the anatomy of the eye, the image of the young man that originally fell on the retina is upside down. And backward. It is sent to the brain that way. The brain's job is to reinvent the image by turning it around and putting it right side up. When that happens, you see something.

Not all of the retina will respond to light, however. Many of you may be familiar with an area in your eyes termed the *blind spot,* technically referred to as the *optic disk.* This blind spot is in a particular part of the retina. All of those retinal nerves busily tattling on the light they have received must collect somewhere and then exit the eye to get to the brain. At the collection point, the nerves are unresponsive to any light that happens to fall on them. The field of view represented by the light falling on that point will not be registered by the brain. This point is thus called the blind spot. In our example, if the light from the fin of the goldfish falls on that spot, you aren't going to see that fin in your vision. It's absolutely correct to say you never really see a certain percentage of your field of view.

At this stage of the discussion, you might point out to me that this is nonsense. If there were blind spots in each eye, we should always have two disk-shaped holes to contend with in our field of view. And we don't have two holes to contend with; therefore, I ought to go back to school or consider another line of work. Since I'm too tired to do one and too old to do the other, I must defend myself with an as-yet-unexplained aspect of cerebral reality.

We don't have two blank spots to contend with in our field of view because we don't see with our eyes; we see with our brains. Your brain "knows" that two blank spots are evident in every field of view. But the brain doesn't reveal the vision of a goldfish-ingesting young man with unexplained holes in either eye's visual field. Instead, the brain gets creative. It takes a look at the disk-shaped hole it received and compares it with the patterns and shapes immediately 360° to the hole. The brain then literally creates an image from the information around the blind spot and fills in the hole with the image it made up. Part of every field of view you perceive, including the page in this book, has been artificially constructed in your brain. That is, your brain makes up a portion of what you see, kind of like a friendly hallucination or a bit of a dream.

Studying the biology of the blind spot demonstrates the power of the brain in determining how we perceive reality. Actually, more spectacular gymnastics are going on between brain and eye. You have to perceive shades of color, you have to integrate the input of two eyes into a single view, and you have to flip and invert every image you see. In a few moments we are going to explore the realm of genetics and human thought. We need to recognize, in doing so, that the ability of the brain to alter and editorialize what we perceive is a powerful aspect of human experience.

The Tragedy of Anencephaly

The end result of a number of congenital deformities (those occurring during uterine development) is the loss of a brain. These deformities are reminiscent of the results of many forms of modern higher education. In the case of newborns, however, the cause is purely genetic.

For example, *acephaly* is a deformity that results in the complete loss of the human head. A normal body is formed except for an incomplete neck stump. This defect is intriguing to the biologist in that it demonstrates the modular aspect of the developmental program. That is, a completely developed lower body is capable of forming inside the womb without the biochemical council of the fetal brain. Once established positionally, different parts of the body develop according to a local and independent genetic recipe. Fortunately, acephaly is a very rare genetic defect.

In other mutations, much more common than acephaly, the head and the brain are only partially developed. Some of these defects seem subtle. For example, a mutation affecting the head at first glance doesn't appear to be a defect at all. The neck, face, and skull are fully present at birth, and the baby appears normal. Except that the baby doesn't make any noise. Or breathe. Or move except in a reflexive manner. When the inside of the skull is examined, the reason for the inactivity is made apparent. The baby is completely missing its brain. In place of the brain is a tumor incapable of supporting human life. The baby, separated from its mother, dies. The disorder, because of the existence of pieces of neuronal tissue, is termed *pseudoencephaly.*

By far the most common disorder is anencephaly. This disorder isn't a single mutation; rather, it is a blanket term referring to a group of mutations with a single characteristic. Here is a formal definition:

> [Anencephalic anatomy includes] . . . the destruction of the brain or the forebrain along with the missing scalp and partial skin defects. Only some neurons and tissue of the choroid plexus, as well as partially necrotic neural tissue infiltrated by angiomatous formations, can be identified in the anencephalic brain. In some cases, the hind brain can be intact. (J. A. Keen, 1962, The morphology of the skull in anencephalic monsters. *S. Afric. J. Lab. and Clin. Med. 8*:1–9.)

What does that mean in English? An anencephalic is a baby with most of its brain destroyed, or, better to say, never fully formed in the first place. That part of the brain still present is naked, covered by neither a skull nor a scalp. In most cases, it looks like someone has taken a giant spoon and scooped out the top of the skull. This mutation seems so hideous because, aside from its head, much of the baby's body remains perfectly formed. An anencephalic can be thought of as a baby born with part of its head, and most of its brain, completely missing.

Anencephaly is hardly a new occurrence in the human race. An Egyptian tomb was uncovered that contained the mummy of an eight-month-old anencephalic baby. In the Middle Ages, such babies were usually referred to as *toad heads,* under the mistaken impression that gazing upon a toad while pregnant could cause the defect. It was also thought that anencephaly could be caused if the mother underwent a sudden and violent shock.

Research is being done on the placement and development of the embryonic head. Recently, a researcher at the University of Washington made a two-headed frog embryo by changing the function of a single gene. But that is a result in the opposite direction from what is observed in anencephaly. And even if it were in the same direction, the result has not been repeated in a mammal. Like so many things in molecular biology, much work needs to be done.

If there is no brain, how long can an anencephalic baby live outside the womb? About one-half of all anencephalics die during pregnancy and are thus classified as stillborn. Because the *hindbrain* (the part of the brain that controls many of the automatic biological functions such as heartbeat and breathing) is present, some anencephalics survive childbirth under their own power. Most stop breathing minutes after delivery, and few survive more than a week without being placed on artificial support systems. Some, however, have survived for much longer periods of time.

The anencephalic is unable to perceive sensations the way you and I usually think of them, even though most of the time it possesses sense organs. For example, the eyes are often fully formed; they just aren't connected to anything. The same thing is true of the ears. And the nose. An anencephalic is blind and deaf and incapable of smelling anything. A normal mouth is present that has a usual, if not slightly thickened, tongue inside it. But there is no way to experience taste because the nerves are not hooked up to their normal switches. The anencephalic is, for the most part, sensorially deprived.

This apparent lack of perception must be contrasted with another capacity of the anencephalic: its eerie ability to react to its environment. Because a little portion of the anencephalic's brain is preserved, it is capable of certain reflexive functions. For example, it has the ability to grasp objects with its hands. It can also exhibit a sucking motion with its mouth. It gasps and gurgles in an attempt to fill its lungs with precious oxygen. It can respond to painful stimulus. This response, however, is not the kind of physical pain we experience when we stub a toe, cut a finger, or recover from major surgery. To feel that kind of pain, the anencephalic would need a part of the brain known as the *thalamus*. Since it does not possess that structure, its reaction is reflexive, possessing the same kind of automaticity we experience when the doctor tests our reflexes with that little silver hammer.

The anencephalic is a mixture of completion and mutation. Of sensitivity and senselessness. Gazing at an anencephalic child has always given me the feeling of a familiar horror, the recognition of untested infancy and the revulsion at the seeming arbitrariness of genetic malfunction.

The anencephalic child is also a collection of tissues that in almost every way appears to be a complete human being. Every way except one: it does not have a brain or at least not much of one. It cannot think in the manner with which you and I are familiar. It is incapable of consciously identifying its parents from itself, of committing a crime, or of embracing a religion. The anencephalic brings the question of appearance as a criterion for sentience to its ultimate test. Is the anencephalic baby, this complex series of gurgling, twitching tissues, also a human being? Or is it more like a tumor? A giant bag of transplantable organs?

As we shall see, the answer to these questions holds the key to the unfathomable happiness of thousands of parents. And the potential millions of dollars in revenue for the biotech industry. As an added bonus, it brings us right to the edge of our question: what we think is biologically unconscious tissue and what we think is vitally responsive life.

Donor Organs

In this section, we will discuss the ideas related to defining the humanity of the anencephalic. We will begin by describing the use of anencephalic children in medicine and biological research. We will then talk about the controversy that such use brings forth: the judgment of humanity based on the presence or viability of the brain. That will mean discussing the biology of human thoughts and what our possession of them means to us. Perhaps science can find a clue to the difference between humanity and tissue by examining the very organ that allows us to ask the question. Or perhaps because we must use our brains to investigate our brains, we will forever be blind to the answer.

Of what use is the anencephalic child to medical research and society in general? It is a good question that could have been answered "not for much" twenty-five years ago. However, recent technologies, from molecular biological manipulations to transplantations, have made the use of the organs inside such deformed babies very attractive. That attraction has provided the philosophical momentum, or the excuse, to place the humanity of the anencephalic in a workable context.

One use for anencephalic tissues derives from the fact that the cells of which they are composed are so young and so robust. Many cells we use in molecular biology behave just like they do in the body; that is, they have a finite life span. If they are from an adult, they usually don't last long. That can be a heartbreaker when you have spent $50,000 studying a cell line which ends up dying on you before you can finish your experiments. Fetal cells can last longer in culture and perform better in both short- and long-term genetic manipulations. In addition, fetal cells are an excellent source of growth factors, molecules that have effects ranging from teaching Aunt Mary how to fight her breast cancer all the way to making tall basketball players. To some scientists, denying access to these tissues would be like cutting down the Brazilian rain forest and losing, arbitrarily, a valuable natural resource.

In addition to research, such cells are used in the treatment of certain human diseases. In a therapy that may one day prove curative, fetal adrenal glands or certain fetal nerve cells have been transplanted into the brains of patients with Parkinson's disease. Fetal bone marrow may also be of tremendous value in fighting off certain genetic disorders or radiation sickness. Because of their extreme youth, these cells may need less immunosuppression after being transplanted; that is, they may last longer in the body of the recipient because that body may not be as eager to reject the newcomers.

Perhaps the largest use for cells derived from anencephalic babies comes from the organ transplant side of medicine. Transplantable organs for those who need them are in short supply. This is as true for newly born babies as it is for not so newly born adults. Hundreds of people die each year because there are not enough available organs to go around. It is one of the greatest heartbreaks in medicine to be with parents who must watch their little ones slowly die because no organ is available for rescue. Especially when the children could be saved if there was a sufficient supply of tissue. In the future, as transplantation technology becomes more widespread and more sophisticated, the need for more organs will only increase.

A False Appearance

The emotional impetus of such events has prompted researchers to look to the anencephalic baby for additional sources of harvestable organs. In many ways, this deformation creates an ideal incubator for various kinds of tissues. Because only parts of the brain are missing, the rest of the anencephalic's body is a virtual supermarket of transplantable organs. Over 2,500 anencephalics are born each year. The tissue is prepackaged and is in a readily available supply. Moreover, it is fresh tissue: anencephalics can be often placed on life support immediately after birth. They would then be dissected while still functioning. If the clinicians waited for the baby to die naturally, certain tissues would be rendered useless to save lives. The anencephalic cannot usually be allowed to expire on its own. Instead, it must be kept on life support until the time of transplantation.

Anencephalic babies bring the argument of appearance-as-life to a disconcerting focus. For the most part, the babies look like living human beings. They possess rudimentary functions of human beings. But they are missing perhaps the one great trait that separates human beings from the rest of the biology, the ability to think. And from that thinking, to derive memory and consciousness. Anencephalics' ability to perceive reality is lost because they have no way of re-creating it in their brains. And no way of filtering it. If a brain is never capable of displaying any of those characteristics, did its owner ever exist as a human?

Dead Center

We have seen from earlier chapters that declaring standard biological life an on/off event is very difficult. So is trying to define the onset of protectable humanity. We encounter the same difficulty when we try to define the onset of biological death. Is determining a definition of human death any easier than determining a definition of human life? The answer is no, which makes discussing rationally an already sensitive subject nearly impossible.

Western culture has deeply embedded into it the idea that there is a moment of death. Somehow this giant, complex organism full of little eddies of independent life and swirls of miniature biochemical theme parks is thought to normally cease functioning all at once. Like a conductor putting down the baton at the end of a symphony, death usually is thought of as absolute biological silence. Is this in reality what happens?

The answer to that question is no. Human death is seldom experienced in a momentous fashion. Usually seconds or minutes or hours or days of biological life are left inside a person even after apparent lethal injury. In recognition of this fact, many well-known people (George Washington and Eleanor Roosevelt, to name two) stipulated in their wills that the blood be removed or the head or the heart be excised prior to the final closing of the coffin.

Other cultures have rejected the notion of death as a sudden on or off proposition altogether. In a number of primitive societies, there is a waiting period after biological life has left before it is finalized in a ritual. In Bali, the waiting period is exactly forty-two days. Certain Malaysian tribes will wait twenty years. The Dyak tribe of Borneo will keep a corpse at home and treat it as a living being for a period of time. It is daily offered breakfast, lunch, dinner, and conversation, and it is constantly surrounded by friends and loved ones. (With the onset of Western public health ideas, the tribe has been persuaded to take the corpse out of the home and place it in specially constructed quarters. There it can be visited and attended to until the burial has been arranged.) In all of these examples, death is thought of as the closing of life. It is a gradual event, only a part of which is the conclusion of biological processes.

Is it really so strange that these cultures see death as an increment rather than as an instant? We might have been tempted to say yes before the onset of the last half of the twentieth century. Modern technology, however, has made it much more difficult for us to define the actual moment of death. So difficult that we may have to change the way we view it.

Consider, for example, the victim of an accidental drowning in freezing waters. There are no vital signs in someone whose body temperature drops below 90° F. There is very little brain activity. The victim appears dead by all modern criteria. In the old days, because of these characteristics, there was no attempt at resuscitation. The person was declared clinically deceased, and a death certificate was filled out. But that doesn't happen anymore. Now it is understood that at those temperatures, the body enters a state of artificially induced hibernation or torpor. These days, great care is taken to revive the person. The body is gradually warmed on the inside and on the outside. After a period of "death," the person springs back to life. For this person, there is no moment of fixed, irretrievable death even though the biological processes that could be measured were shut down. The moment of death existed in the person. But only sort of.

The other side of the coin is examined when one considers persons in comas. Many brain functions persist but very rarely has anyone come out of a coma after many years of being in one. Rare or not, the fact that any come out at all gives reason for hope; their condition may reflect our ignorance of their biology more than the permanence of the condition. Thus, the confusion and the ambiguity surrounding the issues of people suspended in their moment of death exist to this day.

In 1968, an attempt was made to resolve this ambiguity. A group of Harvard clinicians got together and decided to usher in the concept of brain death. The idea is simple. Most nerve cells, when damaged in a particular fashion, are irretrievably lost. The brain, of course, is made of nerve cells. It is also the seat of the intellect and personality and consciousness and identity and everything else. When such an organ is damaged, all the attributes of the person that made him or her uniquely alive are lost. Permanently. Because of that permanence, the death of the person was said to have occurred. At the time, the idea sounded quite straightforward. The only question remaining was that of determining how much brain damage constituted permanent loss.

The criterion these clinicians came up with was the brain wave. And the machine they came up with to measure brain waves was called the *EEG* (thankfully short for *electroencephalograph*). Brain waves can be thought of as the gross national product of all of the electrical activity in the brain at any one time. An EEG measures that electrical output. If a patient was thought to be dead, the clinician could measure the brain waves generated under the person's skull. If there was no spontaneous activity, there would be a flat (negative) EEG; there was no electrical output and thus no life in the person. A death certificate could be filled out and services arranged.

The Unbearable Lightness of Jell-O

Several years after this standard was proposed and accepted, a neurologist in California came up with a drastic improvement on the EEG. He found a way to make it hundreds of times more sensitive than the ones in use around the country; the machine could record events deep inside the brain. Standard EEGs could record events only on the surface. The good doctor used his machine on a number of patients pronounced dead by the old machine. He found that twenty-six people were alive by his more sensitive technology. As a result, several people's lives were saved, and all recovered without any brain damage. The new technology seemed to help finely tune the moment of death and more accurately define who could be saved and who could not.

As with all new inventions, however, things can go awry. Machines can become so sensitive that they measure what they're supposed to measure, and also everything else. A Canadian neurologist suspected that something like this might be happening with the new EEG machine. And perhaps with the whole standard of brain death as well. The physician obtained lime-flavored Jell-O and had it molded into the shape of a human brain. He hooked up the EEG to the Jell-O and started recording. To the amusement, or horror, of some of the staff, he was able to obtain evidence of human life in the Jell-O. Its electrical activity could be measured in full lime flavor, a brain wave, as it were. Now, as I recall, Jell-O is alive only in certain bad Japanese horror movies. The problem was that signals from standard electrical equipment in the vicinity of the aspic were being picked up.

Brain of Terror

The point was well made. And on both sides of the coin. Might someone be kept artificially alive because of background electrical artifacts in the hospital? Yet if the artifacts were eliminated, would a clinician be sure that he or she had not also eliminated a valid signal for life? The line between what is human flesh and what is hospital construction is so difficult to define that in the opinion of a number of clinicians, it is arbitrary.

Such arbitrariness has been played out in the legal system of the United States. Some states have rejected the idea of brain death outright, relying instead on circulatory and respiratory failure as their primary criterion for cessation of life. Some states have made brain death their sole criterion. Others have made circulatory/respiratory failure or brain death the criterion, presumably whichever comes first. So, in the United States, it all depends on where you get your injury as to whether you have died from it or not.

The situation gets worse when one considers the rest of the world. Most countries have adopted brain death definitions. However, if you are in France, the brain wave must be absent for forty-eight hours before a death certificate can be signed. If you are in the Soviet Union, that number shrinks to five minutes. In the United States, for those states that employ brain death statutes, the usual time limit is twenty-four hours. There are no unilaterally accepted definitions about when the moment of death occurs. This lack of standardization shows that it is as hard to define when the moment of death occurs as it is to attempt to define when the moment of life occurs.

Most of these arguments, from the status of a deformed child to the death of a normal adult, have focused on the functionality of the human brain. That is where we are headed as well. To understand such issues, we must examine the biochemistry of the brain activity itself. We shall begin by examining the brains of people like you and me. We must attempt to describe the biology of a human thought.

The Lord of the Dance

I have always been amazed at ballet dancers. The athletic skills and grace such people bring to human motion are almost otherworldly. Such beauty is achieved by continually exercising a body that often resists the thundering delicacy this art demands. The full appreciation of five seconds of graceful motion is realized by understanding that it can take fifteen years to attain it. It's a lot of sweat for a fleeting amount of time. I'm told that in many instances you can tell who has been practicing and who hasn't by looking at the ease with which an artist performs. When one beholds a ballerina in a present production, one is really peering into her past exercises.

My wife and I had the opportunity to witness the Bolshoi Ballet perform Prokofiev's *Ivan the Terrible*. Though it was a profound experience for both of us, we were affected by it for different reasons. Kari positively resonated with the translation of physical movement into the deep emotions of the human experience. Aspects of her thoughts and feelings were being choreographed physically before her eyes; they filled with tears in response. To her, such feelings slipped the gravity of this planet and ushered her into the infinite expressiveness of the divine. Toward the end of the ballet, it was all she could do to keep from weeping.

My reaction to the ballet was not nearly so artistic. I tended to look at the stress and strain demanded of the dancers' joints and ligaments and be amazed that such obvious pain could be displayed so gracefully. Since one of my ongoing laboratory projects deals with genes that make muscles form, I thought of the continual interaction of cells and proteins supervising motion. The ballet was still marvelous to watch; I was deeply grateful that a God who could make bodies so complicated could also make them so beautiful. Nevertheless, each moved for completely different reasons, the two of us experienced the Bolshoi.

We talked about the thoughts and feelings we experienced all the way home. Back at the apartment, I attempted to mimic one move I thought was so grand and promptly fell squarely on my behind. To the subsequent gales of delighted (I trust) laughter from my wife, I dusted myself off and reduced my communication to verbal descriptions. She thought it was interesting and mildly humorous that I could know so much about muscles and so little about movement. The point was expressed very well. Understanding how muscles work does not guarantee that one will automatically become a dancer.

Kari knows a lot about music and ballet because she is a professional composer and pianist. In our discussions about our respective professions, we have come to the conclusion that science and art are sometimes the Siamese twins of the human mind. Science is the twin with the head and is thus heartless; art is the twin with the heart and is thus mindless. This hypothesis has produced no end of, shall I say animated, discussion in our household. We believe that science seems heartless because it cannot produce a qualitative judgment on things it can otherwise thoroughly explain. I have no doubt that a good orthopedic surgeon could also become a great ballerina. But I am aware that her medical training will only get her so far in a dancing career. After that, it's all innate talent and a lot of hard work. In many ways, to place a scientific lid on certain aspects of the human experience is only to describe a part of that experience. And a curiously impersonal one at that.

Is the same limitation true when we talk about the ability of certain biochemistries to produce a quality of humanness? In this section we are going to briefly describe some of the biochemistry that governs human thought. We will necessarily start at the basic unit of any thoughtful nerve-centered impulse: how an individual nerve cell receives and transmits information. We will then examine how that impulse works in situations that are not normally considered to be human thought. The idea is to discover if such knowledge gives us the ability to discern the entire dance of protectable humanity. Or whether such impulses are irrelevant to the ballet taken as a whole and we are merely examining certain muscles.

A Brain in the Neck

Performing a hemorrhoidectomy through the nose is easier than trying to describe the incredible complexity of a single human thought. Human thoughts reside in the brain, an organ we know nearly as much about as we do about successfully designing Hubble space telescopes. This unknown organ is a twisted, writhing mass of nerve cells, and within that seething caldron, human thoughts are created. To understand the subject, we have to talk about how these cells are organized in the brain.

If we were examining the brain like a spy satellite examines the swimming habits of our latest political enemy, we would see distinct and vast sections of brain matter. Two giant hemispheres, collectively called the *cerebrum,* dominate the neural landscape. The *cerebellum,* an area underneath the cerebrum, controls the subconscious movements (whatever that means) of skeletal muscles. Finally, looking a little like the bottom green part of a flowering rose is a supporting structure called the *brain stem.* Deep inside these regions are areas that inform you to eat a Big Mac, fight gorillas, see a ballet, or go to the bathroom. All of these sensations (and many millions more) are capable of being perceived because this massive brain is made of nerve cells (see fig. 8.1).

If this brain was isolated from the rest of the body, the ability to perceive the outside world would be severely curtailed. Fortunately, it is anything but isolated. Hooked up to this teeming mass of tissue, like the freeways that feed Los Angeles, are even more nerve cells. These nerves cascade down the bottom of the brain and fan out to distant points throughout the body. Such nerves have many different functions. Some nerves, among the most valuable in the human collection, ensure that the brain can smell pizza. Others ensure that when your sister's nine-year-old son stomps on your big toe, you know which foot was the target and how hard it was hit. Still other nerves inform you which muscle groups will be required to strangle the little guy if you can catch him.

Nervous Looks

At a gross level, the nervous system can be divided into two parts: (1) the nerves that make up the brain itself, and (2) the nerves that feed information to it. Scientists have made lots of divisions and subdivisions in these nerve systems. But to understand something about human thought, we need to know only that there are nerves that send signals to the brain and there are those that receive those signals in the brain.

That statement implies that nerve cells talk to each other. That's not an

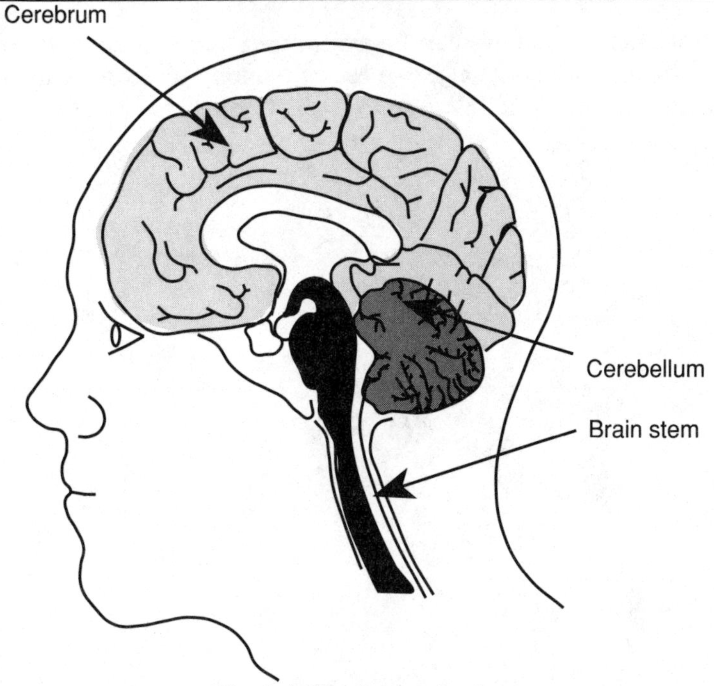

Fig. 8.1. **The human brain**

implication. Nerve cells talk to each other like teenagers on the telephone. One nerve cell will get a signal and transmit that signal to the next neuron. The new nerve will transmit its newly acquired information to the next nerve cell and to the next, all the way up to the brain. And then the brain can toss the signal around its billions of nerve cells like a baseball and send back an answer over a similar route.

This transmission can happen very quickly because the nerve cells lie next to each other, forming long wiry lines throughout the body. Communication occurs only because certain genes in the nucleus of the nerve cells have given those cells some very peculiar shapes. These shape genes, combined with other gene products inside other neural cells, allow a truly senseless brain to understand a truly sense-filled world.

What does a nerve cell look like? And what does that shape have to do with grabbing your sister's little boy and flushing him down the toilet? Nerve cells (if we are to be technical we should call them *neurons*) come in various shapes. Most assume the basic form of an uprooted tree (fig. 8.2). The top nerve fibers, where the branches exist, are the *dendrites*. The long trunk of the tree is an *axon*. The roots, often much smaller than the branches, are known as *teleodendria*. These branches and roots receive and transmit electrical information. The branches of one nerve can intermingle with the roots of another nerve next to it. Somewhere in that intermingling, information can be transferred and subsequently passed on. The shapes of individual nerve cells are important in determining the flow of signals from one nerve to the next throughout the body.

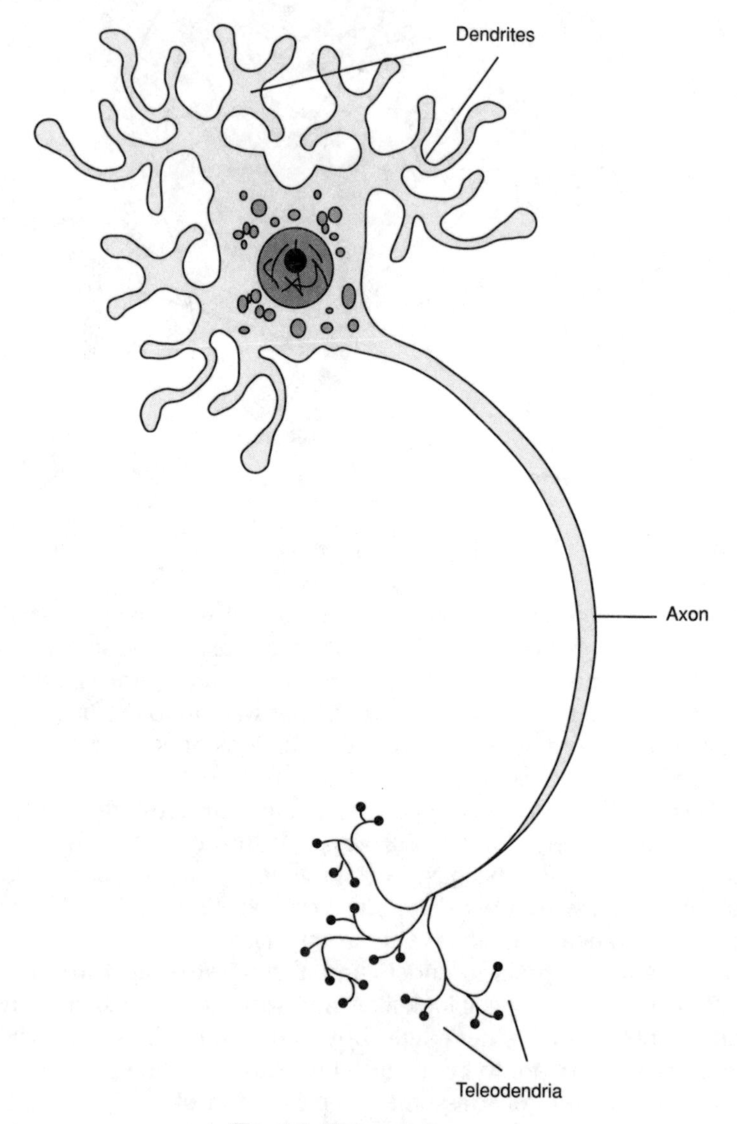

Fig. 8.2. **Nerve cell**

Neurons are not boring you've-seen-one-you've-seen-'em-all-shaped entities. Nerve cells can be as small as a few hundredths of an inch, like cells typically found in the brain. Nerve cells can also be over a yard long,

3.28 feet to be exact. These string beans exist mostly in nerves strung between the toe and the spinal cord. They are easily the longest cells in your body.

Salt and Battery

The business of the treelike neurons, whether long or short, is to pass information from point A to point B. The information that your toe has been stomped is first noticed by the nerves in your toe. What form does the information take? Obviously, there are no little nephews sitting in your toe telling other little nephews residing in your brain that they are all little brats and that one just stepped on your foot. The body needs a way to convert what it sees in the environment into information that the tiny cells in the brain can understand and respond to.

The body possesses such a conversion event. This conversion event is the essence of human thought. Electricity is the currency the body exchanges with external sensations for the brain to perceive them. Any input received from the body is converted into electricity, and that electricity runs right up to the brain. To comprehend how this happens, from the pain you feel in your toe to the anger you feel in your mind, we must talk about how an individual neuron can sense things. And to do that, we have to talk about common table salt.

You know about common table salt, the white stuff you and the MacGregor kid used to put in the sugar bowl to fool your mothers when they would have coffee in the morning and sit and talk to each other instead of you. The salt would turn transparent in the brown liquid, but the obnoxious taste and the spanking you probably received as a result were undeniably real.

You may remember from chemistry class that table salt is made up of two atoms, sodium and chlorine. They are hooked together in a fairly stable crystalline marriage until they hit substances like coffee. The salt will disappear in water because the water destroyed the salt crystal. It literally tore apart the sodium from the chlorine. When the nasty split occurred, the chlorine ripped off a few bits of the sodium atom and kept them for its own use.

In the violent world of solubility, no community property laws are allowed. The sodium atom was left with a debt, a chunk of itself missing, as it were. We call this hole a *charge*. Because we like to think that atoms can profit from being ripped off, we call it a positive charge. So what your mother had floating around in her coffee cup was a bunch of wounded sodium atoms desperately trying to learn from their mistakes by carrying a positive charge.

Getting Out the Volt

What in the world does this have to do with human thinking? A great deal. Your nerve cells, like tainted coffee, are in close contact with sodium. They are bathing in sodium. There is so much sodium near nerve cells that the atom keeps trying to get into the nerve cell. Sometimes they succeed, an event as noxious to a resting neuron as table salt is in Mrs. MacGregor's cup. The nerve cell has to keep kicking the sodium out. It has literally thousands of little molecular pumps on its surface for that purpose. As long as those pumps work, enthusiastic sodium atoms will not be allowed inside the nerve cell.

This sodium, just like its cousins in your mother's coffee cup, is a ripped-off sodium. It is missing some of its own particulate, has incurred the debt, and carries a positive charge. Consequently, lots of positive charges are floating around the outside of a typical nerve cell that is minding its own business. Since it is doing its best to keep the positive charges of the sodium out, the inside of the cell is relatively positive-free. In nerve land, another way of saying that the inside lacks a positive charge is to say that it possesses a negative charge. The inside of a typical nerve cell is thus negatively charged. (To be honest, there are positive charges inside the neuron due to the presence of other atoms—like potassium—in the cell. These are counterbalanced by a relatively large number of protein molecules that are negatively charged. Consequently, much more positivity is outside the cell than inside.) As we'll see, it is critical that the inside of a nerve cell remain negatively charged with respect to its outside. And it's kept that way because, night and day, little sodium pumps are busy finding stray sodium atoms inside a nerve cell and ejecting them (see fig. 8.3). For those of you familiar with electricity, this situation is simply charged with potential.

An Amp Unto My Feet

What does a difference in the location of sodium atoms have to do with the activity of nerves and, thus, human thought? This is how it is all put together. When your nephew steps on your toe, many nerves get stimulated. You already knew that; you felt a great deal of pain and perhaps some revenge when it happened because nerves were stimulated. But let's go down to just one nerve feeling the pressure of that little foot on your big toe. How is that one little nerve reacting?

The answer to that question is "with painful enthusiasm." When the stimulus is received that the toe has been stepped on, the nerve cell encountering the stimulus changes its apartheid policies toward sodium atoms. All of a sudden, the positively charged sodium atoms are allowed to rush into the nerve cell like the Seventh Cavalry, which changes the

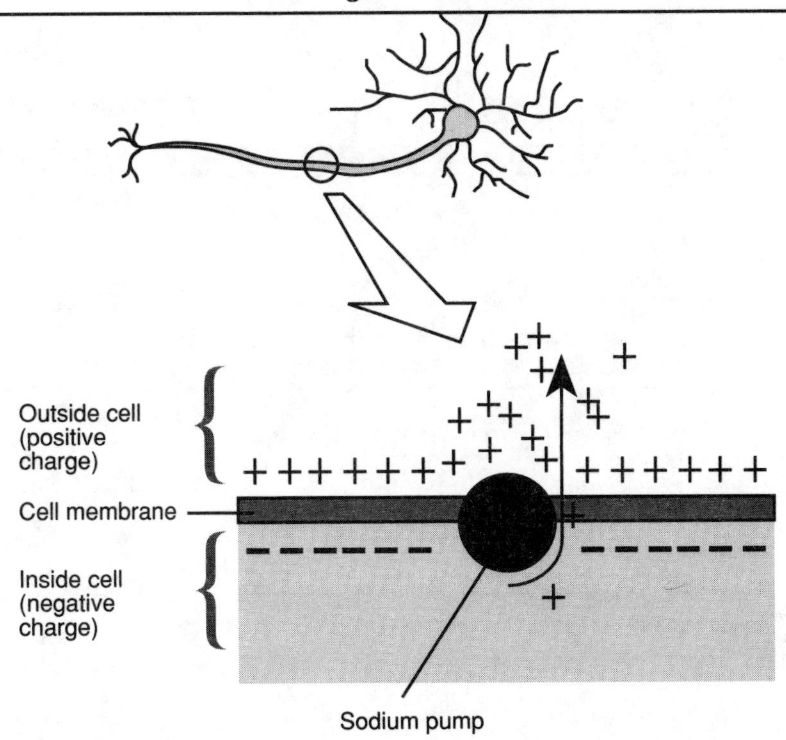

Fig. 8.3. Charges in a nerve cell. Shown is an enlargement of the membrane of a typical nerve cell. The sodium pump works to keep the outside of the membrane positively charged with respect to the inside. This difference in charge location is important in the propagation of human thought.

charge on the inside of the nerve cell. With so many positively charged sodium ions invading the premises, the inside is now positive with respect to its outside. In other words, the charge has moved from the outside to the inside, and those nasty sodium atoms are responsible for the change. All your nephew's foot did to that nerve was to allow a lot of lousy salt atoms inside. But that change in location was a significant event (see fig. 8.4).

The sudden movement of sodium atoms, and I might add quite local change in charge location, sets up what is called in electricity land a *current*, or better, *current flow*. An electrical charge has been generated, something analogous to a tiny bolt of lightning. This bolt is felt throughout the entire cell, and the cell is said to have fired, like a gun that's gone off or a light switch that's been turned on. Whatever analogy is used, the nerve has been activated. In the inexplicable parlance of biologists, it has been *depolarized*. Electricity is generated inside the cell, all from a nonelectric little foot coming down on top of your toe.

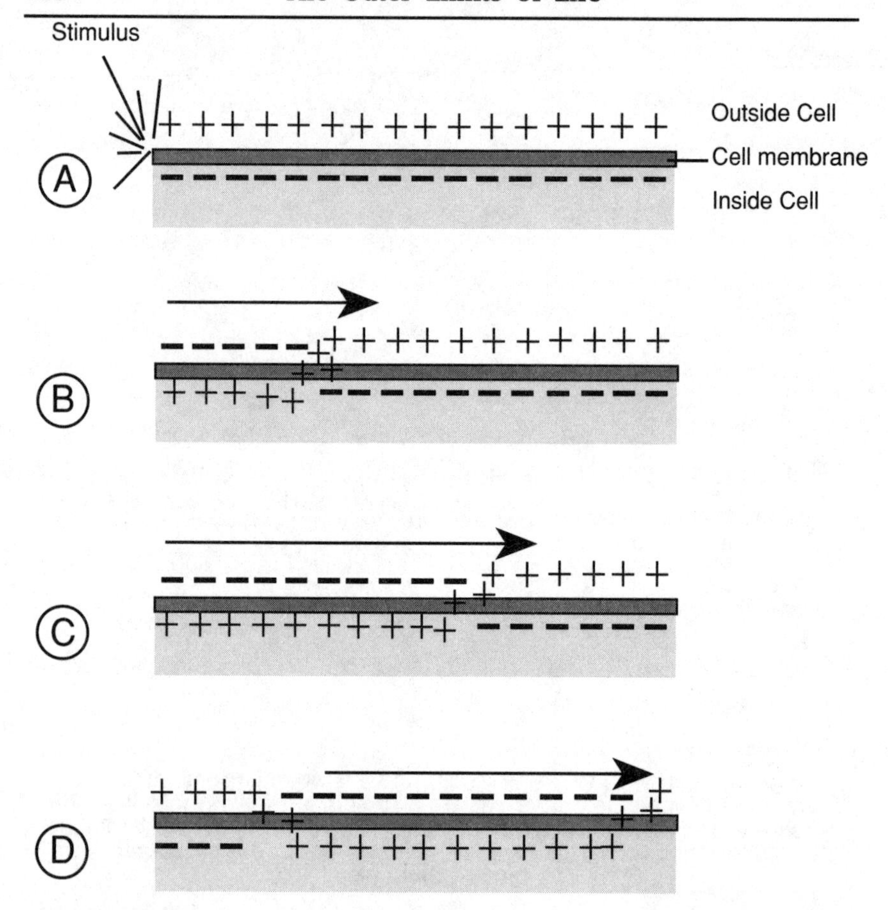

Fig. 8.4. **Propagation of a signal in a nerve cell. A stimulus is received at one end of the nerve (A). This stimulus causes the sodium channels to open, and positive charges come flooding into the interior of the cell (B). This flooding increases in a linear fashion all along the membrane (C). The nerve begins to restore its charge distribution very soon after the stimulus has been received (D). This way, the nerve will be ready for the next stimulus to be received.**

That might be fine for understanding what occurs inside a single neuron in a stimulating situation. Even pretty boring. But if that neuron never got the message any farther than itself, your brain would never know about it. The name of the game with the nervous system is information transfer. This signal must reach your head so that you can feel it and move your other foot out of danger. Before we can conclude this discussion of nerves, we are forced to talk about how this signal gets out of one neuron and into another.

The Convenience of Automatic Transmission

You recall that these uprooted maple tree-looking neurons can come into contact with each other via their branches and roots. The branch of one neuron does not actually touch the root of another neuron, however. There is a little space, a *synapse*, between the ending tip of one neuron and the beginning tip of the other. If there were no way to transfer information across this gulf, this synapse, there would be no way for one nerve to tell another about a certain event.

Fortunately for you and unfortunately for your nephew, neurons have devised a way for information to jump from one neuron to the next. In the end tip of the neuron that first felt the nephew's foot is the molecular equivalent of a navy. A fleet of tiny cargo ships sits listlessly inside the tip of the resting neuron. Identified as *neurotransmitters,* they come in a variety of molecular classes. When a neuron is fired, the last set of molecules to hear about the event is this navy. In response to the electrically charged atmosphere, these molecular ships (literally thousands of them) are suddenly pushed out of the tip and into the gulf separating one nerve from the other. These ships, these neurotransmitters, move across the space between the nerve cells and eventually find a waiting port on the other nerve cell (see fig. 8.5).

Having arrived on the other nerve cell, the molecules dock onto specific sites on its surface. The interesting thing about this docking is the way the new cell reacts to it. The new nerve lets the sodium atoms surrounding it into its interior. The arrival of the molecular navy to this new nerve is also interpreted as a foot being stomped by your nephew. Or at least the previous nerve's response to the stimulus. As a result, this second neuron fires off as if it had been the one stomped on. In this manner, the information of one nerve has been transferred to another nerve.

A Thought in the Dark

So, in a rough manner of speaking, that is the physiology of an active nerve. Is there some kind of organization we can use to make defining thought clearer? We have no real way of breaking into a nerve cell and finding out its favorite color. A single nerve may not be enough. Two examples will illustrate the difficulty of defining a human thought. One is about reflex reactions, specifically regarding the nervous system that governs unconscious thought. The other involves a conscious decision to eat Chocolate Decadence Cake.

Let's say that your nephew did not stomp on your foot. Let's say that you are seated at the dinner table with one leg folded over the other. Sensing a perfect opportunity for mischief, your nephew crawls under the table and gives you a karate chop in the shins. Immediately your leg

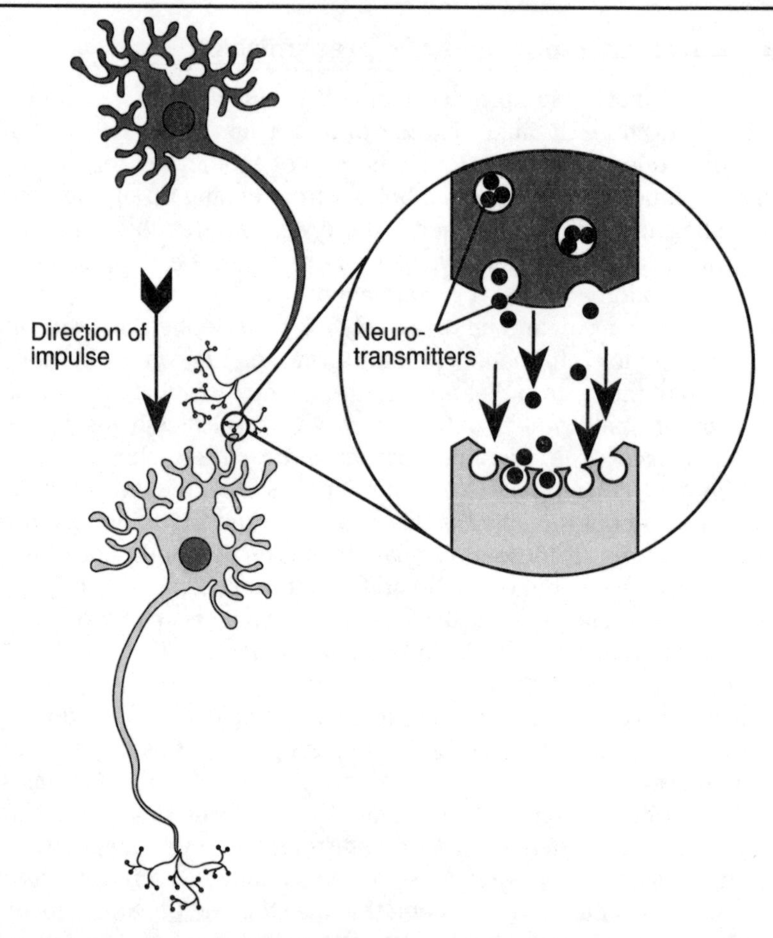

Fig. 8.5. **Enlargement of synapse. As the impulse travels from the top nerve to the bottom, a synapse is encountered. Bags containing neurotransmitters are stimulated to dump their contents into the open space between the two nerves. These transmitters are received by the bottom nerve; as a result, the signal is propagated from one nerve to the next.**

shoots straight out, just as if a physician had hit your knee with a hammer to test your reflexes.

What governed the automatic, apparently out-of-control response? Was that action governed by a human thought, also? Or by little Martians in your kneecap just waiting for karate chop orders from a misbehaving little boy? The answer depends very much on how one defines human thought.

For example, the command saying, "I will now kick my leg out like a Rockette at Radio City Music Hall because of my nephew," was never issued by the brain. The signal to move your leg never got much farther than the base of your spinal column. When your knee nerves felt your nephew's hand, they fired correctly, letting in sodium ions and passing that signal on to the next nerve. But as that signal got passed along, it found that there were no signs saying, "this way to the brain."

Instead, the signal got rerouted onto another nerve at the base of the spinal cord. This new nerve was not headed toward the brain; it was headed back toward the leg. And this nerve had a very different function once it became activated. This nerve would cause the leg to kick into the air.

And that's exactly what you did. The signal formed a *reflex arc*. A nerve stimulus was applied to the knee, sent around to the spinal cord, and directed back to the leg on a nerve with instructions that said, "Kick."

Is a reflex a human thought? And if it is, does this ability to kick determine the nature of human existence? A reflex arc involves human tissue. And it involves activated nerve cells, cells stimulated almost the same way as your brain cells are stimulated when you try to balance your checkbook. Most of you would probably say that the ability to be human involves more than spinal and leg nerves. Especially if you found out that dead people can exhibit the same patellar reflex arc as live people, which in certain situations they can. Most animals exhibit reflexive stimulation, but that does not make them human. So, at its most basic level, humanity cannot simply be the presence of any old stimulatable nerve. There must be a special quality of nerve excitability that makes us different from all the other animals and from each other.

The $64,000 question is, What is that special quality? You can live a full life even if you never get to kick reflexively. What about more vital processes? Like the ability to breathe through human lungs? Or the ability to eat? Do such nerve processes, because of their greater relevance to human life, confer sentience on the owner? Their importance to human life dictates that they be directed by something a little more sophisticated than the back of a spine. That, of course, means the brain. Regions in the brain control processes that never feel like they have to tell your consciousness what they are doing. Are these processes actual thoughts in the same way that deciding to mow the lawn is? To understand if a command for which you are totally unaware constitutes a thought, we must travel directly to fairly mysterious parts of the brain.

'Til Breath Do Us Part

Have you ever sat around on a sunny afternoon and said to yourself, "Why, I think I'll contract my spleen for a little while"? Or gotten cold that evening and verbally directed the blood vessels in your skin to constrict so many millimeters? Or gotten mad at your boss and exclaimed to the pupils in your eyes, "Dilate, you fools, dilate!"?

You haven't? Well, probably neither has anybody else. Those commands fully engage your brain but fully uncouple themselves from your awareness. Consider for a moment your digestive system. Let's say that you have just sat down to partake of a giant piece of Chocolate Decadence Cake.

Unconscious control over bodily processes occurs before you put a piece of the dessert in your mouth. Saliva is being secreted in your mouth all day long. A typical human secretes almost one-fourth gallon of the stuff a day. Now try not to salivate. Put some of that cake in your mouth, and try again not to salivate. You'll find that you can't stop the process. The nerves that control this biochemistry are on automatic pilot, for the most part.

Psychological cues play a role in teaching you to salivate just by thinking of certain foods. Try to stop this artificially learned behavior. Short of cooking something with a pleasing aroma, there's very little you can do to manipulate their secretions because their function is also under automatic control. Nerves have to drink in their sodium and fire so that saliva can be made. But they do so automatically, under control of a part of the brain hidden from conscious manipulation. It does not mean that such places in the brain do not create commands; it simply means you are not aware of it when they do.

With this in mind, let's talk about that Chocolate Decadence Cake. As you eat it, the food that goes down your throat is only inches away from choking you to death. Why is that? The throat you use to pass the cake to your stomach is also the throat you use to breathe. If that's the case, why does the chocolate cake not get into your lungs, clog them, and kill you? The partial answer is that just under your tongue, your throat is divided into two parts. One part goes to your lungs, and one part goes to your stomach. But that's only a partial answer; if you just had a throat partitioned into two sections, half of the time the food in your mouth would go to your lungs, and half of the time it would go to your stomach. And you'd still be in danger of choking to death.

The rest of the answer as to why you don't gag has something to do with a little flap of skin that sits on top of the this-way-to-the-lungs tube. It is called an *epiglottis* and looks something like a leaf. Whenever you breathe, that leaf opens up and allows air inside the tube leading to your lungs. Whenever you eat, that leaf closes over the lung tube and prevents anything from coming inside it. Thus, your cake can bypass the lungs and go safely down into your stomach. When people get something caught in the throat and it "goes the wrong way," a bit of food got into the lung tube before the leaf gate closed. They end up coughing and coughing to expel the foreign object.

The reflexivity of the epiglottal response can be seen by posing the following question: How many of you at breakfast this morning physically told your epiglottis to cover the air tube in your throat because you were sending an omelet to your stomach and you didn't want to choke to death because you had to go to work? Nobody? You are in the majority. Most people don't physically command the epiglottis to do anything. It does its job automatically, continually being stimulated by nerves that get their orders from silent parts of your brain. These nerves suck in sodium and pass their information via neurotransmitters, just like every other nerve in your body. The only difference is that they do it without your knowledge.

Hard-Thought Battles

Many life-support-type functions require no conscious effort on your part to keep you going. Numerous parts of the brain, including the brain stem and parts of the cerebellum, participate in this subterfuge. These areas produce nerve signals every bit as legitimate and necessary as the ones that make it to your awareness.

And that brings us to the central point. Is the definition of human thought broad enough to include these automatic functions? If human thought is defined as activatable nerve tissue, the answer is yes. There is very little physical difference between signals that allow you to select a meal from a restaurant menu and signals that allow you to eat the meal. The only difference is that you are mostly aware of one response and mostly unaware of the other. When electrical activity is measured in the brain, signals emanating from the more automatic centers register just as loud and clear as the ones under more conscious control. This activity represents what is generally termed *living tissue*.

But is awareness enough to separate living tissue from living being? Many anencephalic children possess these automatic pilot centers of the brain intact. They are consequently capable of generating deliberate human thought if we define thought as excitable nerve tissue. They lack the ability to become aware that this is so. Ever. A number of comatose patients are in exactly the same state. Even more head trauma victims, young and old, retain only these most elementary brain functions. They lack only awareness. When is the moment of death real for these people? What is the difference between them and the tissue in a petri dish? Is humanity going to be defined as a continually sliding scale with levels of protectable awareness as the criterion for existence? Is science capable of answering these questions?

Obviously, we need to understand more about human thinking before we address these questions. So far, we have talked about reflexes, the simplest and least accessible kinds of nerve responses. Now we must examine how neural responses come together to produce awareness and intelligence in the parts of the brain in which we do have access. Only then will we see if science is capable of giving us an answer about humanness based purely on neurology.

In many ways these questions remind me of the feelings my wife and I experienced at the Bolshoi Ballet. She can remember the fantastic magic of watching her innermost feelings become transformed to a work of art on stage. And I can be content with watching an enormous biochemical universe become fine-tuned to the precision of a tiny watch. In the end, we both know that understanding human physiology can describe in great detail why a ballet occurs. But to be perfectly honest, we know this understanding is not all that is needed to perform one.

All Thinks Considered

We humans often pride ourselves on having the most sophisticated, highly developed brain of any animal. Which is, of course, why we can build hydrogen bombs or pollute our way off the planet and chimpanzees cannot. Zoologists tell us that we are the undefeated world champion thinkers of all time, even when compared to organisms with twice our brain weight and three times our genetic constituency.

The problem, of course, is that they're right. Sort of. Chimps may not have invented germ warfare, but they also haven't discovered penicillin. Nor have they created symphonies. If you were to look at physiology, that inability might seem a little odd. Ape nerves fire in ways very similar to our own, down to the sodium. But that similarity does nothing to elevate their thought processes to the level of even mentally challenged humans.

The same thing is true of nerve tissues in petri dishes, even if they are human nerves. Isolated from their other connections, they no longer can drive a human to pass the ammunition or praise the Lord. Whether isolation occurs because of birth or dissection, we come to the same conclusion as the last section. What is critical is not that a thought exists but that a certain quality of thought exists. And since this whole book is trying to find the difference between chimps and dishes and you and me, these qualities become extremely important.

This section is dedicated to showing that higher ordered intellectual talents are so hard to identify that they are almost impossible to define, let alone become the reason for human existence. Nerve cells form thoughts based as groups of almost unimaginable complexity. A nerve cell may be hooked up to almost half a million other cells. The connections are so numerous that right now, we can do no better than describe certain vague categories of thinking. Nevertheless, progress has been made, and certain issues are coming into focus as a result of that progress. In that light we will examine two kinds of thinking: memory and consciousness.

We must first perceive what connected, thinking cells look like. You might be surprised to know that a picture has been taken of a thought in progress.

Creating Your Own Reality

Scientists have known for a long time that the brain physically re-creates what the light in the eyes communicates to it. Even as you are reading this book, a picture of this page is being painted with electrical brushes deep inside your brain. Specifically the back of your brain. You can think of your eyes as a movie projector and the back of your brain as a screen on which reality is projected. If certain nerves are destroyed in the back of the brain, a part of visual reality is shut out. However, if the nerves that govern a part of the visual field are destroyed, you don't see giant holes in your vision. No rip in the screen can be visualized. The brain will fill in the blank spots with an image it comes up with on its own or deny that an object within the dysfunctional field exists. This is a repeat of a theme we talked about at the beginning of this chapter: you do not see with your eyes; you see with your brain.

Many scientists believe that in the same way the brain re-creates your visual space, you re-create an action in your brain before you perform it. That is, if you are going to hoist a bale of hay into a truck, you lift up the bale of hay in your mind before you bodily perform the task. This fascinating idea remained in the realm of speculation until the spring of 1990, when an experiment was performed with monkeys that actually took a picture of a thought in motion. The image received showed that an amount of preplanning had occurred before the task was executed.

No Thought for Privacy

The monkeys had a joystick, a lot like the ones used with video games. Before the monkeys was a circular screen with a dim point of light in its center. At a certain time, the dim light would move to the edge of the circle, say, to the twelve o'clock position (if the circle is thought of as a watch face). At this new location, the light would remain dim or would suddenly brighten. If it remained dim at the edge, the monkeys were trained to move the joystick in the direction of the light. In this case, they would move the joystick to the twelve o'clock position from its normally centered position. But if the light reappearing at the edge suddenly became bright, the monkeys were trained to move the joystick 90° counterclockwise to the location of the light. In our example, the monkeys would move the joystick to the nine o'clock position. Thus, the researchers could predict where the monkeys would move the joystick by keeping the light dim or making it suddenly bright.

Next, electrodes, the neurologist's equivalent of video cameras, were hooked up to the monkey's brain. They were placed in nerve cells known to control the motor functions of the animal. The electrodes would allow the researchers to see what nerve cells the monkey commandeered to move the joystick. They would also indicate when the monkey fired those nerves. The researchers then sat the wired-up animal in front of the screen with the joystick, turned on the circular screen complete with its little light, and watched the results.

What happened was amazing. First, they tried the dim light situation, where the light moved from the center to the edge and did not brighten. The monkey moved the joystick in the same direction as the light. They observed a cumulative sum of nerves in the brain firing in the same direction as the movement of light on the screen. And the nerve firing occurred before the monkey actually moved the joystick forward. That is, the monkey reproduced the pattern of movement in its brain and then moved the joystick.

Next, they tried the bright light situation, where the light moved from the center to the edge and then became very bright. The monkey moved the joystick 90° counterclockwise to the direction of the light. The researchers observed nerves in the brain that initially fired in the same direction of the light. But then the firing pattern rotated exactly 90° counterclockwise to the direction of the light. Again, all this happened before the monkey actually moved the joystick forward. The monkey reproduced the pattern of movement in its brain and then moved the joystick.

The experiment was a triumph. It showed an honest-to-goodness thought being played out in the brain of a living animal. The experiment wasn't measuring a behavior already intact in the animal, like an assessment of an instinct. It was a learned behavior in an adult monkey undergoing stimulus/response reactions. The experiment clearly demonstrated that the brain plays out a response before physically initiating it. It showed that the brain maps the physical task it is to perform in a special manner reminiscent of that very task. And most of all, it showed that thoughts are composed of the coordinated actions of many nerve cells, firing in prescribed patterns to perform specific functions.

Thinking As a Social Function

To understand how complex these patterns can be, consider that the average brain contains between 10^{12} and 10^{14} nerve cells. If you lay them end to end and take the legendary let's-wow-them-by-how-big-that-really-is unit of measure known and loved by all scientists, that's enough nerve to circle the globe twenty thousand times. As you can imagine, its packing in the head is extremely dense. If you took out a chunk of brain tissue about the size of a small pea and unwound it, you'd unravel about two miles' worth of nerve cells.

These neural connections form our thinking processes. So far, we have examined the not-so-obvious mechanisms of reflex and involuntary processes. We will now examine two more complex and more familiar talents of the human brain: the ability to remember something and the ability to have consciousness. We will begin with memory.

Chess Pains

Picture a large old hallway, full of the massive shapes and rich, dark browns and golds that reek of early twentieth-century masculinity. The room is thick with cigar smoke and the low murmur of men, some of whom feel they have grasped their culture by the jugular and could squeeze it at any time. This is a men's social club in Edinburgh, Scotland, a local corridor of power that buzzes with the anticipation of pre-World War II Europe.

In the middle of this hallway are thirty-four tables, at which are seated thirty-four men. Spectators surrounding the tables are the source of the murmuring. Each seated man is in front of a partially completed chess game. Opposite each man is an empty chair. The thirty-four men look furtively at the boards and glance nervously at a young man seated some distance away. The young man is playing all thirty-four games simultaneously. And he is playing them blindfolded.

One of the thirty-four men, an older fellow, raises his hand and motions for the young man to come to his table. An attendant rises quickly to assist the young man, and together they make their way. The young man is then seated across from the one who beckoned him. "Bishop to queen's rook 4," the older man says with a rather nervous look. "Black queen takes bishop," the young man says without hesitation. And then he smiles at an old and familiar feeling. "I believe that is checkmate," he says as he rises from his seat. His opponent stares back in bewilderment, and a low whistle is heard, followed by a spontaneous outburst of applause from the spectators. George Koltanowski, age thirty-four, is playing thirty-four different games of chess with thirty-four different men. Blindfolded. And he is winning them all.

In thirteen hours, the games will conclude with twenty-four wins and ten draws. It was an amazing feat in 1937, a record in fact. Even back before intellectual feats were measured, people recognized that George Koltanowski had a marvelous mind. Over the years, other memory masters have come along, demonstrating the extraordinary power of the human brain. We possess these abilities because we can both store pieces of information and retrieve pieces of information. It is a miracle even if you are not George Koltanowski.

Ranks for the Memories

The definition of memory, like that of beauty, gets more elusive as time goes by. Most of us recognize a difference between remembering a friend's face and reciting the pledge of allegiance. There is a difference between remembering how to ride a bicycle and recalling the aroma of coffee in the morning. Moreover, you remember some things only a little while, and you will never be able to forget other things. Memories exist in the brain and are therefore stored as a type of thought. We cannot erase this information if we go to sleep, and we have difficulty changing it no matter how hard we try. Such characteristics naturally incite the curiosity of all of us, including researchers. How do we organize such thoughts? What is the basis for our ability to remember? Does such an amazing ability confer upon us the distinct title of human being?

To answer these questions, we have to explore how the brain works in regard to memory. We must ask questions about the individual nerve cells and how these cells integrate to form an entire picture. And more important, we must understand how nerves store information for future use. To begin with, we must investigate the types of memories we encounter.

Recent research suggests that the brain may have two basic kinds of memories. The first type, *declarative,* remembers such things as "I put red socks in the white wash and got pink sheets" and "Joan has thinner thighs than I do." Most people feel they lose this kind of memory as they get older. That's a correct observation. People who suffer memory loss as a result of brain damage usually lose this kind of memory also. For example, some brain stroke victims cannot remember newly introduced people regardless of how many times they meet them. You can tell a joke, get them to laugh, and tell them the same joke again five minutes later. They will laugh just as hard in its second telling because they have no ability to remember the first time you told it. This is an example of an inability to access declarative memories.

The second kind of memory is known as *procedural.* This is not the memory useful in trivia games. This is the memory of activity. You can ski in the winter even if you haven't skied for eight months because of procedural memory. You can ride a bicycle because of procedural memory. Even if you haven't ridden one in years, you can get on one and, in a few short minutes, be up and running. You may not be able to describe exactly how you do it, but it doesn't seem to matter with procedural memory. The durable recall of the skills is intact and appears to be hard-wired to some motor skill.

A curious feature about memory, declarative or procedural, is its ability to form associations with other things. My wife and I have a friend who always recalls a specific vacation whenever she smells a certain cologne. As a little boy and teenager, I lived in Germany. Whenever I hear the German language, I vividly recall many aspects of my stay in the country. We all have the ability to form memories that associate strongly with other inputs. Often only a hint of a reminder brings a cascade of feelings and remembrances to mind.

Another curious feature about memories is that we don't have many of them within our awareness unless we need them or they are stimulated by an association. Memories are in the attic of the brain and crash into the consciousness only at preinvited signals. With the advent of brain surgery, we have been able to demonstrate certain associations by stimulating a patient's mind with tiny and quite gentle electric probes. Since the brain possesses no pain-feeling nerves, it is possible to put patients under local anesthetic. That way researchers can talk to them, asking for descriptions of what they experience as the probe moves from place to place inside the skull. Smells, visions, feelings, and events are recalled as a result of direct stimulation. The associations are made not by recruiting other nerves but by stimulating the cells directly.

The Internal Reference Librarian

We have learned a lot about memory over the years. Recent evidence indicates that neural associations linked to memory can be broken down into specific categories and that different areas of the brain store different kinds of information, even if the pieces are linked. For example, one stroke victim has a fascinating, though quite unfortunate, brain lesion. If you wrote the word *rhinoceros* on a piece of paper and asked this stroke victim to describe what he saw, he could describe in some detail all that he knew concerning this African animal. He could talk to you about its horn, its hide, and perhaps its nasty disposition. But if you held up a picture of a rhinoceros and asked him to describe what he saw, he could not tell you anything about it. He would not even know what it was. Further tests demonstrated that the patient had a general lack of association between pictures of things and the words describing them.

Combining those test results with other information, the researchers began to realize that the human brain did not always store pictures and text in the same place. Instead, they were stored in different neurons in different parts of the brain. An association was made between the two to recall the full knowledge about certain pictures and the words describing them. In the case of this poor fellow, his brain lesion severed his ability to associate the words he knew with their graphics.

So what does all this mean? Why should memories possess such extraordinary behaviors? Again, it comes down to how the nerves are wired in the brain. The massive crisscrossing allows for greatly sophisticated inputs to come at you from all sides. Just consider that your ears can hear, your nose can smell, and your heart can beat while you are reading this chapter. And if I write *Eiffel Tower,* pictures and memories can flush out of your mind like birds startled in a bush. The neurologist's challenge is to find how these physical organizational patterns turn into the intellectual experiences we feel every second we live.

Phosphates Aren't Just in Detergents

Most of the work exploring the neurological basis of memories has come from animal studies. It has been a source of raging controversy whether animal studies in this area can be applied to higher organisms such as humans. From a molecular biological point of view, the same genes that work in animal brains also work in human ones. Thus, results obtained from other creatures probably have some bearing on what is going on in human systems, even if it serves only as a negative example. Certain experiments in human cells appear to validate this point of view. For our explanation, I'm going to describe memory as if it was happening to you or me. But please, keep in mind that not all researchers agree that such mechanisms exist in our heads.

Let's say you have undergone a specific form of classical conditioning, a lot like Pavlov's dogs learning to salivate for meat at the sound of a bell. Every time you hear a television being turned on, you get an uncontrollable urge to throw a brick at the screen. What goes on in your brain? Let's examine the biochemical mechanism that governs this memorized association.

Neurons or groups of neurons get recruited to remember the association between "L.A. Law" and shattering glass. The recruitment has a specific, biochemical display. Identifying this display is the beginning of understanding how memory works in nerve cells.

In the typical nerve cell is a protein called *protein kinase C,* or PKC for short. Literally thousands of these molecules are free-floating swimmers in the neural cytoplasm. But they are free-floating only as long as they are not required to remember anything. As soon as the nerve is recruited, the PKCs, like aging baby boomers rushing for tickets to the latest Rolling Stones' farewell tour, hustle to the surface of the cell. All of a sudden, thousands of PKCs are bobbing up and down in the cell membrane. This translocation of submerged PKCs to surface PKCs is the beginning of memory storage (see fig. 8.6).

What happens next? These PKCs now grab molecules known as *phosphate groups* and stick them onto other molecules also on the surface of the cell, just like you used to put a sign that said, "Kick me" on the back of

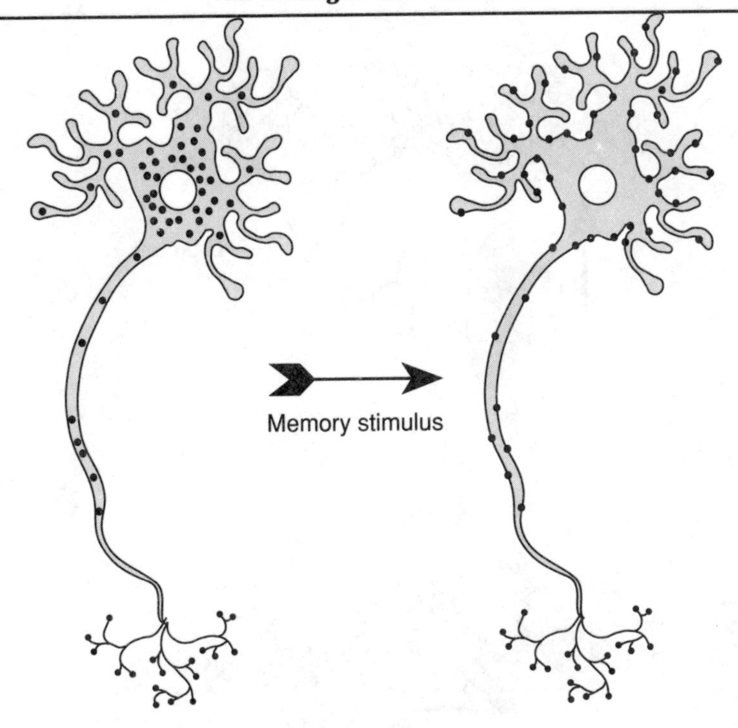

Fig. 8.6. **Movement of protein kinase C. Upon stimulation, molecules of protein kinase C (shown here as tiny black dots) move from the interior of a nerve cell to the surface. This movement is thought to be an important biochemical event in the establishment of memory.**

Suzi Johnson's dress in the second grade. Only in the cell, the target isn't exactly Suzi Johnson; the target is a *G-protein*. This G-protein, in a manner reminiscent of a border patrol, normally decides what gets in and out of the nerve cell. Putting a Suzi Johnsonesque phosphate group on a G-protein changes the immigration policies of the nerve cell. What gets in and out of a particular neuron is altered when PKCs get to the surface. This alteration can be permanent. The change is part of the biochemical process that results (see fig. 8.7) in the establishment of a memory where none was before.

The Changing Shape of Your Brain

The translocation of PKCs to the surface also is a signal for the chromosomes in the nucleus to get with the program and manufacture new gene products. These new gene products can exist for a long time in the nerve's cytoplasm. Their presence may contribute to the long-term nature of memory. These new products may also contribute to one of the

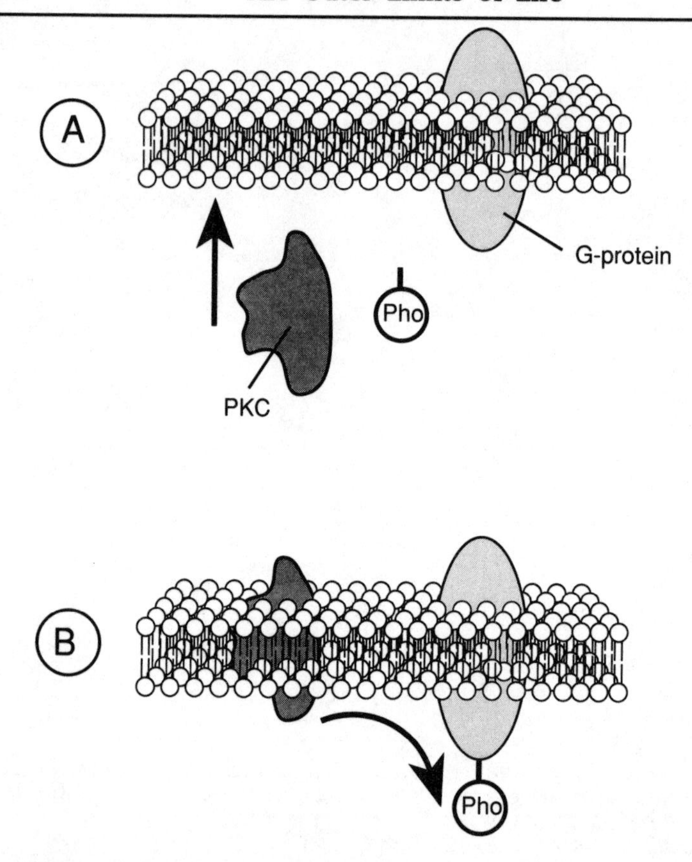

Fig. 8.7. **Biochemistry of memory. PKC is translocated from the cytoplasm to the cell membrane (A). Once in the membrane, PKC mediates the attachment of a phosphate molecule (Pho) to a G-protein (B). This modification results in a change in the kinds of molecules that get in and out of a nerve cell, an alteration probably important in the manufacture of memory.**

most fascinating aspects of the biochemistry of memory—that is, the ability of nerve cells to change their form and rewire their connections in response to something that has been learned.

Does that mean your brain changes shape every time you memorize something? The answer is probably yes. It happens incrementally. You recall that nerve cells learn to talk to each other by the branches of one cell interacting with the roots of the other. There is some pretty strong evidence that nerve cells can change which branch interacts with which root tip when something is learned or can grow new connections and lose other connections. Nerve cells may even move when you memorize something. Or change shape. The result is a fascinating rewiring, simply because of acquired knowledge.

A number of scientists think that multiple memories can be stored as specific connections between two interacting neurons. Since one nerve cell can easily interact with thousands of other nerves, there is a lot of storage capacity. Because you learn something, the internal architecture of the brain physically changes how it looks. These cells have long since stopped dividing, but that hardly means they have stopped living.

So what we have seen in the process of learning to break television screens starts with molecules rising to the surface of cells. This rising causes a change in the ability of nerve cells to import and export certain molecules. It also stimulates genes to make new proteins in the neural cytoplasm. These processes result in a change in the shape and interaction of one nerve cell with another. And what you've got left is a memory of the event.

The next kind of higher order thinking I wish to talk about is consciousness, the ability to be aware of your surroundings and your place in them. That's only a half-hearted definition, however. Self-awareness has so many definitions to so many people that the concept is in danger of becoming diluted to extinction. Neurologists have a tough job trying to understand the biochemical basis for this important, if barely definable, concept. Nonetheless, progress has been made in understanding how the brain is aware of itself in relation to its surroundings. I would now like to relate some of this progress.

Self-Conscious Science

Did you ever sit around during high school and attempt to have profound conversations? Conversations that started out asking if you really existed and then degenerated into discussions about going through puberty in the last half of the twentieth century? My circle of friends in high school often talked like this, the measure of our maturity painfully evident in the memory.

My friend George really liked these questions. He would be fascinated by whether falling trees made noises in the forest if no one was present and would spend hours asking questions like: "Did Adam have a belly button?" Allen, probably the smartest in the group, hated to get involved in such conversations. Dave and I always enjoyed trying to needle poor George, who generally tripped up on his ideas like they were made of shoestrings. For example, George would start the conversation by asking a question like this: "What if I hypothesize I don't know anything?" "That's not a hypothesis," Allen would say. Dave would reply to George, "Well, barf-brains, then you know something." "Huh?" I would ask. "What do you mean by barf-brains?" This would be George, obviously wounded. Annoyed at the all-too-familiar turn of conversation, Allen would exclaim, "He means that if you say you don't know something, you know something. You know that you don't know. Don't you know anything, George?" There would be a brief silence. "Exactly," George would then reply, an inexplicable smirk on his face.

After that, it was time to go back to talking about twentieth-century puberty.

Our first encounters with these ideas were superficial. However, the notion that awareness can be the substance, if not the essence, of existence is not superficial, especially with the advent of certain medical advances. Legally we can call someone dead even if parts of the nervous system are still capable of generating impulses. The individual has lost the ability to be aware of the fact of existence or any other fact for that matter. In terms of defining the road between human thought and human life, consciousness must therefore play a key role. With the unsupported assumption that self-awareness can be measured through Western ideas, the question becomes obvious: What is the biochemical basis for human consciousness?

Having Too Much to Think

We have already obtained a partial hint about biochemical consciousness in our discussion of memory. Much of consciousness is linked to retaining images of events long after the events have passed. Dead persons can't experience memories. If we have consciousness (we may even be asleep), we can experience memories. If we do not have consciousness, we cannot experience memories. Thus, memory and awareness are involved in a marriage to each other for which there can be no divorce.

Even though they are linked, memory and awareness are absolutely different things. Complete amnesiacs are fully able to say, "I know that I am somebody." And you may be able to forget most of the math you learned, but that doesn't necessarily throw you into an identity crisis. If you took the ridiculous worst case, even if you hypothesized that you knew nothing, you would know something. You would know that you knew nothing, thereby contradicting the idea that you knew nothing.

Right, George?

The argument about awareness is old, down to deciding what organ generates it. Historically, finding the seat of consciousness turned into a slug-it-out match between people who thought it resided in the brain and those who thought it resided in the heart. We now know that the heart is a glorious pump but not much more than that. All the action resides in the brain. But the question about the consistency of awareness is not answered just because we know where it resides. What exactly is consciousness? When does a brain stop becoming bits of dark gray-and-white meat and start becoming you and me? And what does this have to do with deciding what is human tissue and what is human life?

Racing the Level of Awareness

One way to get at a biochemical definition of consciousness is to begin simply, like asking when you become aware of a task you are performing. Research has begun to show some interesting results in regards to awareness. From simple tests like moving the hand, some rather startling findings have emerged. For example, a signal is always generated in the brain just before you decide to move one of your hands. It is called the *readiness potential*. About a half second before your hand moves, your brain fires up this readiness potential like a signaling system.

Nothing is really surprising about this result. If the brain truly is the CEO of all thinking, you'd expect it to order such a signal before your hand moved. What is really odd, however, is the answer to this question: When is the exact moment you became aware of deciding to move in response to a command to move? You are aware of deciding to move your hand two-tenths of a second *after* the readiness potential fired. You heard the command to move your hand, your readiness potential fired as a result of the command, and you became aware of deciding to move the hand. It's like your brain automatically decided to respond to the command and then you became aware of it. Further, in the time it normally takes for you to give permission to move your hand, you can just as easily veto the decision. But you are only aware of the veto after that readiness potential has fired.

The most logical question to ask as a result of these findings is, How could that be? Are higher order parts of the brain as much outside your control as your inability to control digestion? Is your brain an automatic robot with consciousness playing the role of executive decision maker?

Many researchers are interpreting this result as a real live biological window on consciousness. They believe that the intention to act arises from brain activity that is not in our conscious awareness. Consciousness instead acts as a gatekeeper or final inspector over decisions and options the brain is dreaming up all the time, seemingly on its own. In this view, consciousness is nothing more than a filter of automatic intentions continually generated by the brain. Actions are executed, or not executed, only when they pass inspection. I should probably point out that there is some controversy concerning the interpretation of this idea.

Prior Knowledge

Other examples show a separation of the brain's stimulation from its ability to be aware of it. One kind of brain damage changes the person's ability to pay attention to certain things. There are patients, for instance, who can't respond to any stimulus that originates on the left side. They have reduced eye movements to that side. They draw pictures that ignore the left sides of things. Such patients fail to dress the left side of the body. In a chronic disorder reminiscent of the John Birch Society, they have lost their ability to understand any of the activity occurring in the world on their left.

Here is the surprising finding about these patients. They can be shown a bright light in their left field of view. Asked if they saw anything, they will respond in the negative. The patients can be hooked up to a machine that will measure their brain activity in response to light. They can then be shown that same flash of light in their left visual field. The machine easily records that the brain responded to the light, just as it would in uninjured persons. When the brain-damaged patients are asked once again if they saw anything, once again the answer is negative. The brain responded to the light, but the patients were unaware of the response. This particular brain lesion resulted in the persons' inability to be aware of what they were seeing.

Another example of awareness versus response comes from patients suffering from *prosopagnosia.* It is also a symptom among stroke victims. These patients cannot recognize faces, not even familiar faces like those of spouses or children. In an experiment to measure cognitive awareness, several patients suffering from prosopagnosia were hooked up to an electrocardiograph (EKG), a machine that measures heart rate. Their spouses or children were ushered into the room, and the patients were asked if any of them looked familiar. The EKG recorded that the patients' heartbeat greatly accelerated as soon as familiar loved ones appeared. Yet when asked for the identity of the people, the patients responded as they always responded: they said they had never seen the people before in their lives. At some level, the patients obviously understood that they were recognizing people they cared very much for. That's why their hearts reacted. But the recognition signals were not able to reach a place where the patients could consciously perceive them.

Such research has begun to break down the process of awareness into component parts. There is a response part; the brain reacts to a command to move a hand, see a bright light, or recognize a loved one. And a part of the brain brings to our awareness that we actually perceived those signals. That awareness gives us the ability to know how we are reacting to our environment and to make decisions based on those reactions. It shows us that seeing something and perceiving it are separable events. It also demonstrates the disconcerting fact that fully functioning individuals do not always have access to their mental functions.

The discovery of this access problem brings up something of an identity crisis. Which process decides what consciousness is? Is personhood the ability to respond to a command or the ability to filter it? How does one define personhood when faced with only portions of awareness? Both involve electrical activity, and both therefore are governed by the genes in the cells. Each has an important role in the overall functioning of a human being. But deciding what is human when only part of these areas is functioning is extremely difficult.

Summarizing These Thoughts

The presence of anencephalic children brings up an extremely unpleasant topic, the death of a human being. Most of us would like to define death, if we want to think about it at all, in terms of a specific moment. We would like to think that an exact point in time exists when an individual's biochemistry completely and irreversibly shuts down. Death often appears that way, whether it's a fictional Gary Cooper being shot in a movie or a very real friend involved in a head-on collision. However, the presence of anencephalic children demonstrates that moments of death do not exist for all human beings. There is no grayer area in all of medicine than the shadowy difference between those that are not really living and those that are not really dead.

Because of the nature of anencephaly, and indeed many neurological deficits, the brain's electrical activity has become the central fulcrum upon which we decide its human sentience. The same standard is used for normal borns as well; whole brain death is a criterion for whole human death throughout the world—even if it is an inconsistent criterion. The problem with anencephalics is that not all of the brain is dead. There's enough electrical activity in many such babies to sustain independent survival for a period of time. Harvesting their organs while brain activity registers is a clear violation of most whole brain death statutes in the United States. And that's where we run into a mighty conflict. If anencephalics are to be useful for transplantation on a large scale, they cannot be allowed to die on their own; in most cases, too much organ degeneration occurs. But to stay within the limits of the law, medical personnel can't harvest their organs until they die on their own. Which makes for a circular argument the Department of Defense would envy.

All these issues depend on how human thought is wired to human existence, which entangles us like flies in a spider's web in our original question. This chapter has dealt with the biochemistry of human thinking in an effort to clarify the connection between nerves and life. As usual, science has not clarified anything but only increased the number and availability of ambiguous questions.

The first thing we encounter about human thinking is that nerves fire. This firing produces electrical activity that can be passed from one nerve to the next. But is this activity enough to define sentience? Not in all cases. A lot of electrical activity is devoted to functions of which we aren't even aware, like eating Chocolate Decadence Cake or breathing. Reflex and involuntary reactions are the only kinds of thought an anencephalic baby is capable of generating. The anencephalic is missing the ability to be aware of itself and its environment. That inability throws the question of sentience down a very greasy intellectual hill. Is awareness, or potential for awareness, the criterion for sentience? To understand that question, we needed more biological information than the mechanisms that drive reflexes and digestion. We thus examined two higher order brain functions, memory and consciousness, to see if their biochemistries gave us better insight into nerves and life.

Renewing Old Memories

First, we found the same thing about memories that we found about reflexes. They complicated our definitions of what we assumed was human thinking. The thinking processes that drive memories could not be isolated to a single nerve whose firing we were perfectly aware of and could predict at will. Instead, a memory apparently could exist in a little branch of one nerve cell or be part of a vast network of many cells, quite scattered throughout the brain. Second, the existence of certain memories was not always within our awareness. We became aware of specific reminiscences only when associations brought them to mind. Thus, most memories are outside our normal consciousness. That we can access them is true, whether by living in a foreign country or by surgery. But that the nerves driving our memories do not fire all the time is equally, and thankfully, also true.

How, then, do memories and our awareness of them affect our humanity? Do we gain more sentience if we can remember more things? Do we lose sentience if we cannot remember a certain percentage of stored facts at any one time? Anencephalic babies don't remember much. Neither do we if we don't sleep for seventy-two hours. Yet we are freely considering making an organ farm out of one population and using them at will for the other. What's wrong? Mostly that we are trying to view a complex process with incomplete data. Memories are not irrelevant in coming up with a definition of human thinking. Neither are they sufficient to describe the whole story.

Conscious Decisions

We also considered the biochemistry of consciousness in attempting to define human thought. The first fact we determined is that we could not get any clearer definition of these processes than we got with memory. But the startling finding suggests that we, like Descartes, may have been on the wrong track altogether. Descartes always assumed that his brain would define his identity. The research into awareness demonstrates that individuals do not necessarily have direct access to their mental functions. Thus, the controls for awareness, like those for digestion, may be beyond our awareness. What is human thinking under these conditions? Do we have to be aware of a cognitive process for it to be a thought? Just how much awareness must we dial into a person before we will call him a human being? Should Descartes change his famous saying to "I think, therefore I am partially out of control"?

The real way to demonstrate the fragility of our definitions of life based on thoughts is to actively consider what happens to them in the hospital. I am referring to the use of machines that measure brain electrical activity in the emergency room. These EEGs can in no way distinguish individually which cells are active and which ones are not. They have a hard time with lime Jell-O. These machines really don't tell you if sentience exists except when there is an absolutely flat (negative) response. And if you measure anencephalic children, a flat response is not the signal you'll receive. Thus, they are hardly discriminating enough to illuminate for us the road between nerves and life. The fact is illustrated quite nicely with the comments made by the people who first invented the brain death standard. Asked when the moment of death occurs as measured by the machines, the inventors answered, "it's an arbitrary decision."

It is difficult to exactly define human thoughts primarily because of incomplete information. We do not know enough about neurology to know when the brain will organize a sneeze, fall in love, or take out the garbage. We do not know which nerves will be chosen, or when we will be aware, once the brain has made a decision. The connectedness of these cells ensures that we will not fully comprehend those processes for a long time to come. So, we cannot understand the fruit of their organization. We cannot define specific thoughts in biochemical terms. Indeed, we may not be able to define them in physical terms at all. It is interesting that we pride ourselves on the ability of our thoughts to separate us from animals, and yet we do not most of the time know what our thoughts are.

Having to constantly rework these arguments can be frustrating for a clinician. Physicians have to do something with brain dead patients, and transplant surgeons would just love to save lives with available organs. The problem in coming up with a definition of the moment of death is coming up with a definition of human thought. It's a critical issue because how we define thinking directly affects what we will call life. But science doesn't give a pat answer. And if there is anything we want from complex questions, it is a pat answer.

So what do we do with our question? In many ways, it is a matter of how we look at it. We have said from the beginning of the chapter that we do not see with our eyes just because they collect light. Instead, we see with our brains, which integrate the wavelengths our eyes collect into coherent pictures. One reason why defining a moment of death is tough is that we mistake science for those integrating brains. We expect science to fill in every blind spot we uncover. That's a shame. And also a problem. Science does not have the power to integrate what we see because it is not the brain. God has given us little bits of light out in creation. Science, like an eye, is capable only of collecting them.

Chapter 9

Smart Imitating Life

Introduction

When I was growing up, we had a backyard neighbor who had a backyard fence built in an unsuccessful attempt to contain his backyard dog. The canine, appropriately named King, was not just an ordinary dog. King was a cross between a German shepherd and the entire offensive front line of the Chicago Bears plus injured reserves. He was also part airplane. No enclosure built during the cold war of the early sixties could have restricted the movement of the dog if he felt the urge to move.

And he felt that urge quite a bit. Every Saturday morning—at 8:00 A.M. exactly—King would bound over his urban enclosure and establish a beachhead in ours. King would march over to our garage and begin a general assault on the contents of our garbage can. The precise maneuver consisted of the surgical removal of the top of the steel receptacle, followed by a large crash as it hurtled to the cement floor.

The next sound came from a human source, emanating from our house in a voice we partially heard and partially felt. It was the slow rumble of my father's vocal cords as painful memories of distant Saturdays slowly defrosted his sleepy brain and dripped onto his waking consciousness. With an angry war whoop he rocketed out of bed and ran out onto the porch. He always arrived just in time to see, but not prevent, the next series of events. The dog pulled everything possible out of our garbage can and displayed its contents on our porch like a garage sale for landfill developers. The only conduit left for Dad's angry energy was a lecture about canine euthanasia, which he promptly marched over to the neighbor's to deliver.

There came a time when Dad got fed up with sabotaged Saturday mornings. He decided that human intelligence, when pitted against canine indigence, would win a resounding victory. And so he proceeded to formulate a plan to exact revenge, a plan that consisted of attaching an electrical device to the entire garbage can. The can was rigged so that only the curiosity of a wet canine nose would be required to send off a mild, but extremely uncomfortable shock to its owner. He set it up on a Friday night in victorious anticipation of waking Saturday morning to the painful howl of a startled, though (hopefully) uninjured dog. We all thought the idea was great fun, especially my brother and I, since we always had to clean up the mess. Next morning, the entire family, like ghoulish Romans at a gladiator fight, got up to witness the spectacle.

We arrived at our breakfast table—which gave a great view of our yard —at 7:55 A.M. in preparation for King's usual 8:00 A.M. invasion time. We waited five excruciatingly long minutes. Unfortunately, there was no dog at 8:00 A.M. Not even a bark. We waited another excruciatingly long sixty minutes. At 9:00 A.M. there was no dog. At 10:00 A.M. there was no dog. Mom gave up the watch and decided to fix breakfast. We finished breakfast by 11:00 A.M., and there was still no dog. Not until almost noon did that dog come bounding over the fence. But he did not have garbage inspection on his mind.

We figured that his owner must have fed him early. King was not interested in eating. Instead, he was interested in securing his territory by the ancient canine-watering ceremony. So we watched him as he set about watering our roses, Mom's garden, our fence posts, and our house corner. Finally, he went over to our garbage can. Then I noticed a small smile creeping over my dad's face. That dog lifted his leg to mark the familiar garbage can in his own aqueous image. And the smile on Dad's face turned into a large grin.

You do not have to be an electrical engineer to know that when canine territorial fluids hit amazing magic electro-cans, the owner of the fluids completes a mighty circuit. A very personal circuit. When King's watery trademark hit the can, he experienced a shock that had a very intimate port of entry. In response, the poor dog let out a howl that rivaled a football stadium's cheer for a home team touchdown. His cranial neurons ablaze, his reproductive future in question, the poor dog leapt across the fence as fast as his legs could carry him.

King never again entered our backyard. He gave a hundred-yard berth to our house whenever he traveled through the neighborhood. After we moved away, our neighbor wrote to tell us that King died without ever setting foot on our property again.

That dog, in no uncertain terms, understood the meaning of electric shocks. Since my brother and I no longer had to clean up a messy backyard, so did we. In that circumstance, humans and dog had memory and consciousness at certain levels. But there was a difference. We knew that the garbage can would never be wired again. The dog knew only that he didn't want to test the idea. He generalized the scene of the injury and, as a result, wouldn't go near the house. Clearly, the animal could think. And because of that capacity, he learned a lesson that would stay with him the rest of his life.

Science is a process that can be just as comical as preventing a dog from trashing a backyard. And sometimes science is a process as mysterious as a bolt out of the blue. Our attitude toward it all depends on what is being studied. The engines that govern our cognitive abilities are processes that at times appear comical and at times infinitely impenetrable. We know as little about what made that dog think as we do about the thought processes Dad employed to wire the garbage can. There certainly are differences between human and animal thinking abilities. But exactly what they are remains something of a mystery. The uncertainty makes it much more tempting to walk around the perimeter of the whole issue of thought processes than to confront it. To be honest, at this point in our technology, it may be all we are capable of doing.

The purpose of this chapter is to explore what is known about the biology of some of those differences. We must talk about intelligence, specifically the difference between animal and human intelligence. Then we must talk about the onset of such a difference in the developing fetus. This exploration is undertaken because we may have found a difference between tissues and sentience we can chew on. That is, an animal is a thinking organism but is not a human; a human is a thinking organism even though an animal. What is the difference in thinking between the dog that learns about an electrical garbage can and the human who rigs it up? Is it the same difference that separates human tissues from human beings? Do we finally have an issue that science can clear up for us, making a moral decision so obvious there is little room for alternatives?

Life should be so simple.

A Brain on Our Resources

As scientists, we are often afraid of a moral question, especially when it taps on our door and asks to come in. It is intriguing, in this light, to read about the guilt experienced by scientists involved in the creation of the first nuclear weapon. Molecular biology is no less powerful, especially when the implications of certain experiments apply for emigration across bioethical borders.

In the last chapter, we talked about the biochemical mechanisms of human thought in an attempt to define it. We also discussed that scientists have a hard time coming up with meaningful descriptions for a phenomenon that at present is mostly undefinable. That's fine; scientists are not omniscient. We may be able to answer it in time, we may never answer it, or we may find that the question is irrelevant. The problem is that certain moral questions, whose satisfaction depends on such a definition, are not tapping on our door. They are battering the house down.

In this section we will explore the differences between animal and human intelligence. We will attempt, as a result of the contrast, to arrive at a working definition for human brain power, which may be easier than trying to describe a human thought. But before we get started, a summary of the latest up-to-the-minute research on this comparative cognitive biology can be stated with ease.

Somewhere in our brains, we know there are differences.

That's the most comprehensive summary. We can confidently affirm what is known about animal versus human intelligence only because so little is clear. The biochemistry of any kind of animal intelligence has been worked out only in a very few organisms. The biochemistry of any kind of human intelligence has not been worked out at all. We know only to ask the question because the differences between humans and plants, excluding the last decade of college undergraduates, are so obvious. However, the mechanisms guiding a bird in flight and a bee in motion are different, too. That does not mean we understand the biology of either.

A Chimp on Your Shoulder

Even though the distinctions between animal thoughts and their human counterparts are vague, we have based our treatment of them on the differences we observe. For example, it is legal for a researcher to chop the head off a monkey. It is illegal for a researcher to chop the head off his department chairman. What is the difference? One is an animal, and one is a human. We have created a double standard in regard to the worth of biological life. We have subordinated the worth of the life of an animal to our own curiosity. We have elevated the life of a human beyond it. We have conspicuously drawn a line in the moral dirt and have challenged all biological life to step over it. Other than our own species, no animal has had the smarts to make it across.

Why did we create that double standard? Obviously, we perceived a difference between human sentience and animal sentience. We know that there are no animal journalists, with the possible exception of the White House Press Corps. No apes compose novels. It's not that a chimpanzee writes badly and a human writes well. A chimp doesn't write at all.

Because of this lack of intellectual talent, there is no Serengeti School for the Gifted Baboon. We can't define a progression of human cognitive talent in most animals. For example, we cannot find a spider monkey that was a baroque composer and compare it to the more advanced gorilla that was, perhaps, a neoclassicist. An orangutan cannot come up with a biological experiment and have a human carry it to completion. This ability to think, to draw analogies, to do whatever one defines as human intelligence, is a function that exists nowhere else in the biosphere except in us. It may be the height of biological chauvinism, but in the end, God has allowed us to stand on a real peak. We cannot discuss the concept of intelligence without treating ourselves differently from the rest of the animals on our planet.

Numerous experiments have been done that drive this point home. First performed almost a half century ago, they have been repeated at various times under various conditions. The model has been to raise animal and human infants in exactly the same environment (to the extent that this is possible) and discern when the cognitive processes begin to differ. A newborn human baby is introduced to a family unit (either by adoption or by parental birth) at the same time that a newborn chimp baby is introduced. They live under the same roof. They are diapered the same way. The stimulus, learning, and attention given to one are given in equal measure (as far as that is possible) to the other. The baby chimp goes wherever the parents would normally take the baby human. Other measures are taken to ensure that the simian baby is treated not as a pet but as a member of the household. During the course of the experiment, both infants are subjected to cognitive and motor tests. Records are kept on the mental and physical progress of each one, usually accompanied by a journal for personal observation and reflection.

How do the baby chimp and the baby human compare? Initially, the chimpanzee does very well. In the first year of life, a chimp matures at roughly double the rate of the human child. The ape exhibits greater muscular coordination. He can tell where sounds originate (aural localization) and can hear noises at much lower decibels than the human baby can. The ape learns to eat with a spoon and drink from a glass before his human counterpart does. The baby chimpanzee shows greater comprehension, cooperation, and obedience to commands. The ape has a better memory and learns to announce his bowel and bladder needs first.

But slowly, a geneticist would say inevitably, the differences emerge. The human learns to manipulate small objects with increasing dexterity and skill. He learns to imitate, draw analogies, and generalize information at a level that leaves the chimp in the dust. The child gradually acquires speech, which provides a window to his environment unlike any other in the animal kingdom. No respiratory barrier prevents a chimp from speaking; the larynx, pharynx, tongue, and facial construction are all present. There is only a cognitive barrier reflected somewhat in the anatomy of his brain. Although a chimp can learn hand gestures, he doesn't naturally use any words (phonemes combined into morphemes), and the ability cannot be instilled by training. A firestorm of controversy rages about whether the learned gestures are simple imitations acquired by the ape's watching his trainer.

Eventually, the reason becomes obvious why a human child can grow up to understand calculus and an ape cannot even say the word. The human learns cognitive skills that far surpass the skills of the brightest apes, regardless of the learning conditions employed. I'm a firm believer that environmental factors contribute more to cognitive development than most of us realize. In the end, however, one cannot fight the pronouncements of the genes. The human has a brain that can teach him how to get along in society, fight wars, discover pharmaceuticals, and learn the distance between heaven and hell. The ape has a brain that does not allow him to write his name.

The Claims People Play

Is the ability to understand these differences cognitive chauvinism? You bet it is; and we end up looking good if for no other reason than that we do the measuring. The problem is that it also happens to be fact. The difference lies in the ability of the respective brains to comprehend and interact in the environment. Whether we look good or not, this difference allows a border to be established between human and animal thoughts. We gain a foothold on human intelligence even if we are studying what it is not.

Does this demarcation help in defining human intelligence? Perhaps just a little, and with some limitations. Although borders may be useful in an overall view, they do nothing to define what lies inside them. We cannot find out where our house exists simply because we know it is not in Antarctica. Which brings us back to our original question: Exactly what is human intelligence? In the next section, we will address this question in two ways: first, by looking at how intelligence is measured, and second, by looking at the brain of a very famous person. We will have an external view, looking at tests that attempt to peer into intelligence from the outside. And we will have an internal view, looking at an anatomy which attempts to define intelligence from the inside. Perhaps by looking at the relative contributions of both perspectives, we can fully appreciate the incredible chasm between them.

Classified Material

So how does one define human intelligence? This question is a little like trying to ask what color a mirror is. It all depends on the image being reflected off its surface. Similarly, the answer that a brain reflects to the listener depends on the kind of question being asked of it. And no single answer is the true color of human intelligence. For example, there is common sense, and there is street sense. Some people are creative geniuses; others have minds that might as well be etched onto computer chips. There are those who are musically gifted, those who are good at language, and those who have excellent people skills. Some individuals are natural athletes; some are brilliant artists. How does one provide a framework for a definition of intelligence that can include all of these talents?

The answer is that you really can't put intelligence into categories with any biochemical meaning. At least not yet. That doesn't mean people haven't tried. One researcher proposed more than a half century ago that intelligence was divided into 2 categories, general abilities and special abilities. Another proposed in 1967 that intelligence was divided into 120 categories. In 1971, a researcher said we needed to go back to 2 categories. A recent suggestion is that there are 4 categories of intelligence. In the future, the number of categories to describe intelligence will be decided by holding a state lottery.

Why has there been such apparent discrepancy? In defining intelligence, articulating and separating individual talents is complicated. Most of us know when we meet smart persons, but pinning down exactly what is smart about them can be very difficult. Do they have a good memory? Are they quick? Do they have tape on their glasses? Some people who can play twenty-three orchestral instruments can't write a complete sentence, and other people who can write in twenty-three languages can't carry a tune. Someone who is a composer of music may also be good at math. At the same time, brilliant physicists may be inept at following directions to a friend's home. How does one distinguish these various gifts in a meaningful way? Is that the correct question?

Uninvited Tests

The answers to those questions await further biological research. And by turning to genetics, we don't clarify very much. We have all heard of the so-called autistic/savant/retarded children who possess incredible memory skills. You tell them the date in the future, and they can tell you the day of the week it falls on. A number of skills are like that, all coming from brains we have thought to be severely intellectually challenged. Developmentally challenged children might hear a complex piece of music once and play it back instantly, note for note. Yet these same children may not be able to write their name on a piece of paper. There are those who can recall incredibly trivial things, yet cannot multiply seven by nine. These facts have caused researchers to think of intelligence in terms of independent modules. The idea is that if one unit is destroyed or impaired, it doesn't necessarily affect the functioning of the other units.

This, of course, is reminiscent of the descriptions of intelligence in terms of categories, an idea that has hardly needed encouragement.

Attempts to define intelligence have been based on more than the presence or absence of certain talents, however. One measure used to define intelligence has been the familiar IQ (Intelligence Quotient) test. First developed in 1905 as a way to distinguish retarded French children from nonretarded French children, the test has grown to an almost bewildering array of examinations. As the number, complexity, and applications of IQ tests have grown, the interpretations of the results have become increasingly controversial. Some persons hail these tests as psychology's greatest achievement; others hail them as psychology's greatest joke.

It is generally agreed, regardless of one's feelings about them, that the tests examine only a small subset of human intellectual capabilities. Moreover, they tend to test only what a child or an adult has learned rather than innate raw horsepower. An IQ test thus becomes an achievement test or an exercise in taking subsequent tests.

Anyone who has taken an achievement test knows that how well you score can be a function of the circumstances under which the test is given. The same is true of IQ, which can vary twenty to thirty points depending on the conditions at the time of administration. For example, one child's IQ reportedly dropped fifty points after the death of his mother. Another troubled child's IQ was raised thirty points after a period of counseling.

The reason for this variability is that your IQ score is not like your shoe size in terms of relative constancy. You don't get a solid number of IQ points at birth like you get eye color. This test has proved useful only in measuring how well someone will perform on IQ tests. From a research scientist's perspective, this test has not proved useful in measuring someone's native intelligence.

Seeing the Light

Is there any way to obtain a definition of intelligence beyond observing that humans have it and eggplants do not? It once again depends on what you are measuring. With administered tests, the measurement depends on a subjective response from the person. You'd like to have something more automatic, like a knee reflex. Or the universal love of chocolate. Something that depends only on the presence of a biochemical process rather than on somebody's mood or attitude.

Attempts have been made to define a variable that would accurately reflect an internal intelligence. For example, increased electrical activity due to a certain stimulus indicates that the brain is doing something in response to that stimulus. Like every time you see someone you have a crush on, your brain goes crazy, and your heart rate accelerates. Or whenever you see a bright light, your brain bats its electrical eyelids. Perhaps measuring how fast a brain can respond to such stimuli would give a clue as to how smart it is.

In an experiment only a space agency could create, researchers at a university in Missouri and a group at National Aeronautics and Space Administration (NASA) studied the response in the brains of twelve-year-old girls to bright flashes of light. The idea was to measure the high- and low-energy electrical activity in girls with high IQs and those with low IQs. So, electrodes were hooked up to the scalps of girls exhibiting a wide variety of intelligences. Every time a light was flashed, the brain response was recorded.

This experiment actually demonstrated something. The girls with high IQs had lots of high-energy responses to the light, much more than the girls with lower IQs. Conversely, the girls with low IQs had lots of low-energy responses to the light, much more than the girls considered smarter. The comparison between these two responses yields an intriguing peek into the world of human intelligence. Unfortunately, it is an indirect peek. In the end, it may not say anything at all. Since we don't know what IQ measures, we don't really know how the observed electrical response tells us anything about the biochemistry of brain function. The results represent a toehold on something, but we don't know if we are clinging to the side of a mountain or to the side of a four-poster bed.

This experiment seeks to define intelligence based on function. It certainly isn't the only way to ask physical questions about intelligence. Another way is to examine the brains of geniuses and compare what you see with the brains of normal people. This approach is based on the time-honored assumption that if something looks different, it must be different.

The Architecture of Intelligence

To perform such visual experiments, one must visit the local medical school, interrupt the gross anatomy lab, and get some brain tissue. I have always thought that gross anatomy lab was well named, especially for people who have never seen such a lab, like many first-year medical students. Their reactions upon seeing their first cadaver indicate they think so, too. *Gross anatomy* refers to the overall anatomical shape and composition of human organs and their relation to one another in the body. It is axiomatic in biology that much can be learned about a creature's abilities by studying the shape and composition of the organs. Human anatomy is no exception.

Many have attempted to find clues to the biochemistry of human thinking by looking at the gross anatomy of the brain itself. The idea has borne much fruit. Researchers have discovered that the brain can be mapped;

that is, certain regions of the brain are responsible for certain intellectual processes. For example, a region of the brain is responsible for speech. Another region processes visual cues. Still another is the center of abstract thinking. A curious by-product of this mapping has been the discovery that the brain contains cells called *glial cells* that have no obvious function. There are about ten times more glial cells in the brain than nerve cells, and nobody knows why. I will have more to say about glial cells in the next section.

Researchers have also looked for regions or structures or patterns in brain architecture that might give a clue to the basis of human intelligence. The best experiment is to compare smart brains with not-so-smart brains. Or perhaps examine a truly exceptional genius and see if there is anything truly exceptional about the brain's internal anatomy.

One intriguing investigation has centered on the brain of Albert Einstein, who died in 1955. The legacy of how his brain got out of its cranium and into a researchers' lab has been the subject of speculation for years. What is known is that the preserved brain was found in a doctor's office in a small town in Missouri. Dr. Marian C. Diamond was given the opportunity to dissect and study the brain of this great man.

The direction for the research was guided, in part, by Einstein's description of how he solved problems. Einstein said that his most fruitful thinking depended on finding relationships between signs and images. That would mean examining a region of the brain known to be associated with certain intellectual neural functions. It is called the *association cortex.* This association cortex does not receive direct stimuli from any of the senses, like vision or smell or touch. Instead, this region collects input from far-flung locations in specific areas of the brain.

Dr. Diamond looked at this cortex in a part of the brain known as the parietal lobe (see fig. 9.1). She found an interesting neural architecture in this part of Einstein's brain. In a normal person, there are not very many glial cells in this lobe; the ratio of glial cells to neural cells is about two to one. However, in Einstein's brain, the ratio of these cells to each other was about one to one in this area. Thus, the numbers of these kinds of cells Einstein had available were different from what is normally seen. And they were different in an area of the brain where higher orders of thinking are important. Thus, like the bright light/IQ experiment, a positive correlation was observed between intelligence and a physically measurable phenomenon.

But does the examination of this area of Einstein's brain say anything about human intelligence? It would be nice if Dr. Diamond could have found the neurological equivalent of a Rosetta stone in Einstein's brain. Unfortunately, she couldn't. The ratio of glial to neural cells, even from the mind of a great man, tells us very little about human intelligence. We can only examine the fact that such differences exist and hope in future

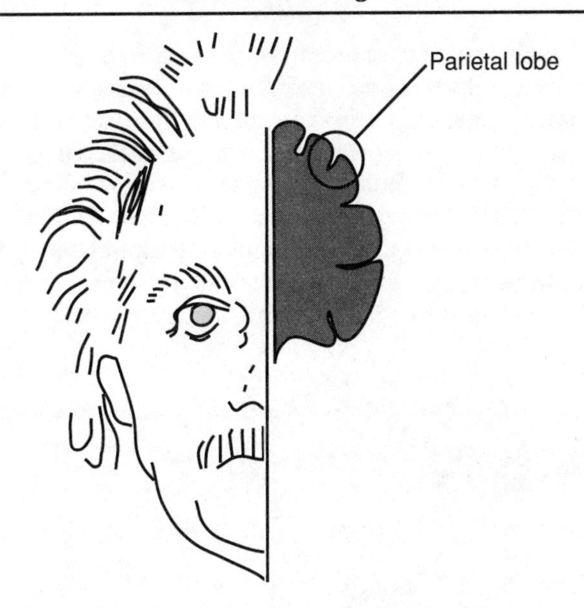

***Fig. 9.1.* The examination of Einstein's brain. Dr. Diamond looked in the association cortex of the parietal lobe.**

decades that the insight is valuable. Human intelligence, so tantalizingly obvious, unfortunately remains just as tantalizingly inaccessible.

Not Settling Our Differences

So what do we have here? We have a phenomenon that is extremely obvious and, at the same time, extremely out of reach. Like the Big Dipper. Or the wisdom of God. At this point in our technology, we can make only vague comparisons between processes that are very different. In terms of measurement, we know some people perform better on certain tests than others do. This fact has been resident in our minds since the first grade. Not even measurable biochemical facts, like the response to flashes of light and the neural interior designs of great minds, tell us much about intelligence. We only know that it is there, like Mount Everest.

Thus, the jury is still out in regard to the decision about the definition of human intelligence. In fact, the jury is on vacation. Which means that attempting to extract an overarching definition of human life based on a biochemical definition of human intelligence is a waste of time. Or at least premature, except in the vaguest sense. Since we are dealing with real live flesh-and-blood humans, being vague is probably not a good platform from which to launch giant moral decisions.

And then again, being vague may not be entirely useless. Our inability to climb Mount Everest does not mean that the mountain has ceased to exist. And since we start out as unintelligible fertilized eggs that can't tie our shoelaces and end up as intelligible adults who can, we vaguely know when intelligence parachutes into our skulls. Can we find out exactly when it landed, even if we don't know what arrived? Or is finding the change just as nebulous as finding the definition, simply because they are unalterably linked? Obviously, understanding the biology of brain formation in the womb would help. And so would attempting to discern when the linkages capable of creating human thought become resident within a human embryo. In the next section, that is exactly what we'll do.

The Magic Fruit

My first experience with genetically manipulating organisms occurred when I was three years old. It was a hot, breezy summer's day in eastern Washington State, the kind of dry heat that feels like God washed the air and decided to blow dry it with the wind. My mother, having fixed us a wonderfully cool little outdoors lunch, went back inside the house to fix dessert. A few minutes later she brought a giant green football-like object, dripping with water. I didn't think that anything that shape could possibly be eaten without cooking it first. Especially when she cut it open and revealed a red, moist interior. So I failed to heed the urging to take a bite until she told me that it was a fruit, it had been sitting in the refrigerator all day, and it was cool, and if I didn't eat it, my brother Dave could have it all to himself and then what would be left for me? That deprivation would be too much to bear. I eagerly sank my teeth into the piece offered me in an effort to catch up to Dave, who was already munching contentedly on his. My mother knew that I would probably like it a lot and so looked on with amused predictability as my eyes lighted up after the first bite of watermelon.

Liking the taste of watermelon hardly constitutes a genetic experiment, however. The idea for the research came when Mom told me toward the end of the summer that the watermelon season would soon be over and wouldn't return until next year. Tears in my eyes, I demanded to know immediately where the watermelons were going. That led to a discussion about where watermelons came from in the first place, which led to a discussion about the little black seeds I wasn't supposed to swallow but ended up doing so anyway because my brother Dave did.

My mother gently told me that if next summer I were to find a garden, moisten some dirt with water, add those little black seeds, and wait, watermelons would grow. I was severely disappointed with her answer because we didn't have a garden. But I reasoned that if you needed only dirt, water, and seeds, you could grow watermelons anywhere. Since I didn't want to wait a year to eat the fruit, I figured the best way to have them around would be to try to grow them out of my mouth. That's when the research experiment began. I drank some water, ate two handfuls of melon seed, and then stuffed as much dirt in my mouth as I possibly could.

I immediately noticed something was terribly wrong with my idea.

Two hours later, writhing in pain in the emergency room and grief-stricken that instant watermelons would not be a part of my life, I watched the disgusting contents of my failed experiment yield to the physician's stomach pump. I learned a great lesson of physical research science that day: scientific maturity is not a gift conferred on someone like a medal. Instead, it is a muscle gradually developed after relentless cognitive exercise. And by the passage of time. That lesson was to be extended into graduate school, where I was to find that research becomes a faculty member. A hard one; she always gives the test first and teaches the lesson after. I am told she conducts a classroom that you never really leave.

In this section, we are going to explore a process that is every bit as gradual and slow to develop as learning to conduct a good experiment. We are going to talk about the development of the human brain, specifically the biology that underpins this unique human organ. To do that, we will first talk about the biology of the nerves and then discuss their intellectual capabilities at various stages. The hope is that even if we can't define exactly what intelligence is, we might be able to discern when we possess the ability to display it. And perhaps find directions in the map showing us the place where the sentience exists in our cells.

Gene Sprouts

How is the brain of a normal individual formed? By eating lots of Twinkies, watching thousands of hours of TV, and trying to figure out demon-possessed income tax forms written by people who have learned about our native tongue from a book called *English As She Is Spoke*? Of course not. That is how a normal brain *learns* things. To understand how we develop the capacity to get mad at the federal government, we have to review a few facts. We have to go back to the first month of life postconception.

You might remember that when the embryo is about three weeks old, a little canal forms down its back. Under the microscope, this canal looks like someone put a finger in the embryo, as if it were made of clay, and formed a soft indentation down the center. You might also remember that this little canal submerges into the back of the embryo. As it submerges, the canal folds onto itself until it forms a tube. That tube, once fully inside the embryo, will begin to form the spinal column. And at one end of that tube, sprouting like a small flower on a long stem, will be the brain.

So how does this intellectual little blossom unfold to form the brain? The action takes place on one end of the tube, which has a bulge at its tip (see fig. 9.2). This bulge has an outside, and it also has an inner lining of cells, huddling together in a little knot like a football team. All the action takes place in that inner lining of cells. New cells are created from this inner lining and start moving to the outside of the bulge. The bulge gets bigger. And bigger. As the bulge grows, new waves of cells move to the outside and pile on top of their brethren, like the football team had just tackled a quarterback. Only instead of the growing mound becoming a random pile, the cells are laid down in a very specific order. That ordering is extremely important. It has allowed our species to hurl rockets to distant planets, reach deep into our genetics, and create the concept of income tax.

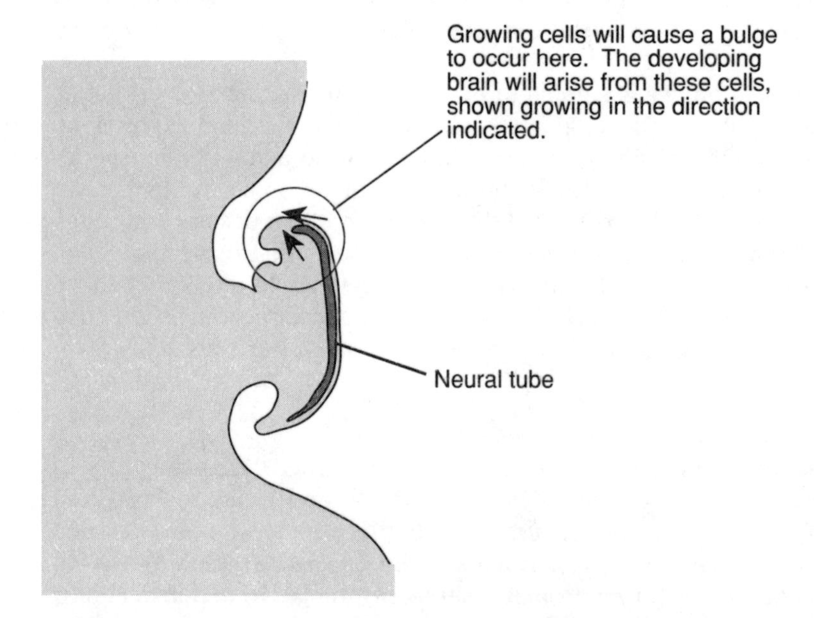

***Fig. 9.2.* The beginnings of the human brain. The embryo shown is about twenty-three days old.**

Because this organization is responsible for whatever intelligence we have, we need to discuss how the cells know where to go to develop in the brain. Or where not to develop. Two sets of processes allow this organization to take place. One set involves making the equivalent of an amusement park ride. The other involves the manufacture of a love potion.

A Cranial Matterhorn

The cells that begin this guided migration out of the tip of the spinal column do so only because certain cells went out beforehand and made a pathway. Cells called *astrocytes* are a subset of the glial cells we talked about earlier. Named after a well-known space-age cartoon dog, these cells boldly go from that inner cell lining onto the outer surface of the bulge before the brain starts forming.

You can think of astrocytes more like the construction crew in an amusement park. The crew has a very specific job to do: to build a giant cellular mountain. But not just any mountain. These cells are genetically programmed to build a mountain filled with tunnels and holes and passages and underground trails of almost unimaginable complexity. Much like a mountain at an amusement park.

The cells do not secrete substances equivalent to cement or stone or wood planks to build this mountain. Instead, the cells themselves become the mountain. It is best described as a living mound. The interaction of these cells with each other forms the passageways and tunnels and inner chambers of the structure. Eventually, there is a vast number of these cells. In an adult, much of the brain is not made of excitable nerve cells; a lot of it is made of these mountain-building cells.

What is the purpose of building an amusement park ride at the top of the spinal column? It's done for a very specific purpose. You recall that when the real nerve cells come out of their huddle in the inner cell lining, they start piling on top of each other. Actually, they try to pile up on each other. As they leave the inner cell lining and go to the outer bulge, they immediately encounter this labyrinth of cellular passageways under construction. These traveling nerve cells, if they are going to make it to the top, are forced to move through the developing tunnels and passageways (see fig. 9.3). Their movement depends on the presence of this scaffolding. Thousands of nerve cells stream into these tunnels and holes and chambers. As the nerves traverse the mountain, the bulge at the tip of the spinal column gets bigger and bigger. Presumably, more tunnels are formed at the top of the mountain as more nerve cells enter the passageways from below. The cells that have most recently entered the bottom travel all the way to the top. Thus, the younger cells are always at the outer edge of the growing brain; the older cells are deeper inside.

A 24-Karat Mold

Forming the brain in this fashion guarantees that it will come out in a specific shape. And that it will have a very specific internal structure. It's kind of like the Jell-O mold approach to building organs. It also means that the genetic information needed to make much of the brain's architec-

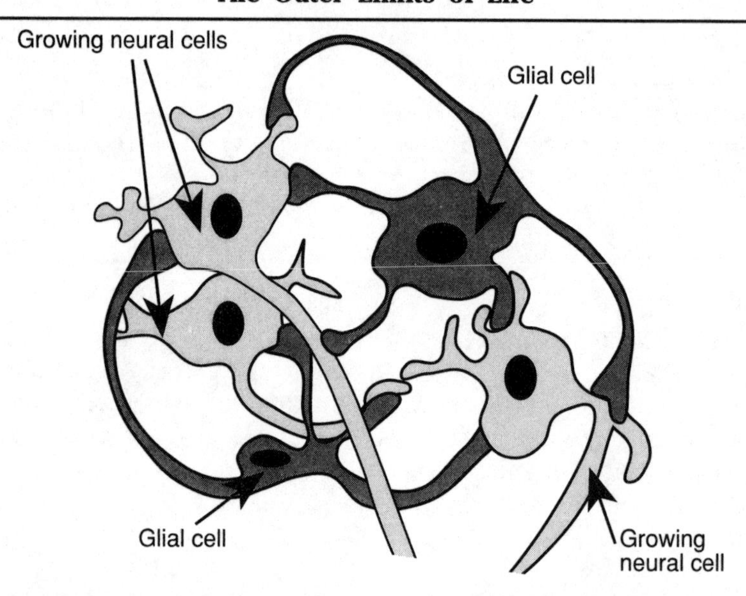

Fig. 9.3. **Highly schematized view of the relationship between glial cells and growing neural cells. Glial cells are thought to provide the scaffolding that supports neural cell placement and growth. The patterns formed between interconnecting glial and developing neural cells comprise the interior three-dimensional architecture of the brain.**

ture is formed by cells that will do none of its thinking. These astrocytes will play critically functional roles beyond just mortar and brick. But for the moment, we can think of brain formation as an interaction between the cells that form the tunnels and the cells that slither through them.

Sometimes a Great Potion

Creating a cognitive Matterhorn and wiring up a brain capable of climbing the real one are two different things. The scaffolding provided by the astrocytes makes available the structural platform for nerve cells to meet and form specific little groups. But the individual nerve cells must get together and form the connections. It's a little bit like having the pastor build a giant hall so that a bunch of junior-highers in the youth group can get together and fellowship. He can build the building, but it is up to the kids to meet one another.

How do the nerves get together? Now that they have occupied this mountainous giganto funplex, how do they form the millions of connections we see under the microscope? The answer teaches us a lesson in neural socialization. Nerve cells are not particularly shy about meeting one another. Different nerve cells will grow and try to connect with other nerve cells. Some will form connections, and some will not. The question in brain organization is, What makes nerve cell A connect with another nerve cell and nerve cell B not connect at all?

Understanding the answer to that question may be as mysterious as attempting to understand why one human falls in love with another. In the case of brain manufacture, it may have something to do with the synthesis of certain chemicals that different nerve cells make. It is thought that genes in nerve cells produce proteins called *trophic factors*. These trophic factors can be thought of as a cross between a love potion and a Sunday meal. If the right nerve cell comes along, it might be immediately attracted to a nerve cell secreting a certain trophic factor. As a result, the two nerve cells will connect, the newcomer eats dinner, and grows. But if the wrong nerve cell comes along, there is no attraction and thus no connection. It is not fed dinner and eventually starves to death. This is a cruel kind of courtship. If the nerve is Mr. Right, he is invited in to make a permanent household. But if the nerve is Mr. Wrong, he does not get the connection, and he dies.

A Balance of Error

What we have in brain development is an equilibrium between the same two forces that impinge on most human experiences, love and death. The question is, What makes one nerve cell Mr. Right and one nerve cell Mr. Wrong? The answer, for the most part, is completely unknown. Different response genes are probably being expressed in different nerve cells. Some may be genetically wired to respond to Love Potion #9, and some may be able to respond only to Love Potion #4. There is a biochemical matchmaking service going on, with selection being decided on the basis of fortuitious—or perhaps deliberate—location. We end up encountering lots of well-connected nerve cells and lots of dead ones in the world of the developing brain.

All of this architecture happens in a specifically timed, if poorly understood, sequence of events, somewhat like Little League baseball. Nerves start forming around the fourth week after conception has occurred. They aren't greatly connected, and they won't be able to form anything resembling an organ until those Mountain Amusement Park cells get their orders to build the scaffold. That anything that could be called intelligence has begun to form in the first month of life is extremely unlikely (see fig. 9.4).

Who's Calling the Thoughts?

By the end of the twelfth week, the beginnings of this mountain have formed, and the nerves have long since invaded. The nerve cells are sufficiently connected that a brain wave, separate from the mother, can

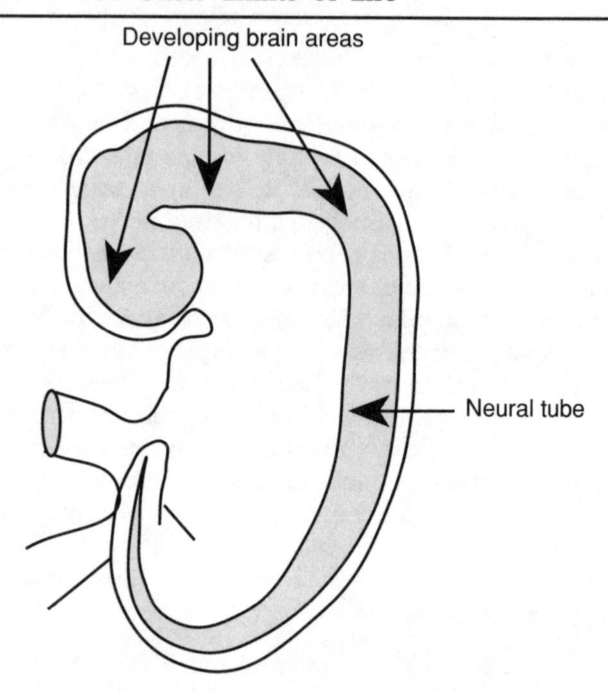

Fig. 9.4. **The developing human brain (shown about one month postconception)**

be detected. What does this mean in terms of the kind of intelligence available to the fetus? We only know about actions. Reflexes are available for the fetus. It can yawn and suck its thumb and hiccup. It can also swallow and has begun breathing motions. The fetus is doing these things on its own power. These movements are spontaneous instead of reactions to external stimulation. So one must take the term *reflex* with a grain of salt here. The nerves are beginning to flex their wings, so to speak, in intentional, though one might say random, actions. They are not yet a series of coordinated actions under the control of an all-knowing cerebral cortex. But they are displaying deliberate activity.

So when do higher order functions, like inner-experience perception (consciousness, pain, and so on), occur in the developing fetus? Obviously, these experiences will not happen to the fetus until those sections of the brain responsible for them are built. Which means that the minimum age a fetus could experience such higher order functions is about twenty-five weeks, assuming that it depends on higher order functions at all. What is so special about twenty-five weeks? At this time two very important structures, the thalamus and cerebral cortex, formally annex themselves. This union means that the electrical activity of the brain will be much more coordinated. After the merger, the resultant brain waves are similar to the adult counterparts in terms of sleeping and waking. From here on, the fetus has the cognitive biological equipment for higher thinking. If human intelligence can be played across these nerves like a bow on a string, the fetus has sheet music at this time.

But does it really? Is anything there simply because of the presence of structure? Or is there something else that must kick start that group of nerves into intelligence? No one really knows. The kind of IQ test that must be administered to measure such intelligence has yet to be invented. It really isn't clear that inner experience depends on a significant degree of brain function. At this stage of our knowledge, self-awareness may exist, even if it is only unresponsive listening. Thus, certain kinds of honest-to-goodness nonanimal human intelligence could occur way before the twenty-fifth week. Or not occur for many weeks after that date. That we are beginning to understand the order of neural structure does not mean we understand the order of neural function. These questions are frustratingly difficult, mostly because many lives are riding on their answers. Any judgments we make now on so little data could show that we have insights far beyond our scientific maturity. Or show that we are making a great mistake and have become as absurd as a little boy I once knew, trying to use his mouth as a place to grow watermelons.

Gland Over Fist

It wasn't as if Mike's mom was an emotional woman. Yet if your son had not communicated with you in years and then suddenly you heard him ask how you were, you might get a little bleary-eyed, too. As it was, she rushed into the living room in response to his question and stopped short at a familiar sight. There was her son, strapped upright in a wheelchair, wires springing from his head like a punk rock haircut, staring straight ahead. Only this time, his usual vacancy was not nearly so painful. The wires had pulled something out of Mike's head and into his mother's ears. She sat down beside his wheelchair, put her head on his lap, and began to sob.

Mike's mother was hardly the only one in the room with tears. Several scientists from the University of Illinois and from Philadelphia were in the room, too. And they were rejoicing and crying. One of them recalls when he heard the words spoken, he went out onto the front lawn and did cartwheels for a while. You might not think it was so spectacular that a loving son, a college honor student and former navy officer, would ask his mother how she was. Mike had a lot of healthy friendships and a good relationship with his parents—all the attributes of a productive, well-adjusted person. Mike had a good mind all right. The only thing Mike was missing was a functional body that could take orders from it.

The scientists were in the living room because they were performing an experiment in human communication. Doctors described Mike's brain as being locked in. He possessed a fully functional, alert mind cemented inside a completely paralyzed body. An automobile accident years ago left him disconnected from most of his human functions. Even his face, including mouth and vocal cords, was not operational. The scientists wanted to see if they could get Mike's thoughts out of his brain and into the rest of the world so that he could communicate with it. And it with him.

How were they going to accomplish this feat? What kind of machine could possibly read thoughts and reveal them to the outside world? What does this technology have to do with human intelligence? Mike's ability to ask his mother how she was doing has a great deal to do with the topic of this section. We are going to discuss the incredible flexibility of the human brain. We are going to explore the unique ability of the brain to rewire itself and adapt to very specific circumstances. We will explore Mike's brain and also our own, specifically those regions involved in sight, hearing, and speech. We must explore this facet of our gray-and-white matter to gain a clearer understanding of human intelligence. And to understand the issues involved in determining when intelligence is conferred on a group of cells.

But back to Mike. The goal was to use the electrical activity in his brain to trigger a computer that could read his mind and display what it saw to the world. But we can't really appreciate what happened to him until we talk a little bit about brain function, specifically certain brain waves. We will begin by describing a very particular kind of electrical activity.

Let's imagine that you have just called a good friend on the phone. You dialed her phone number minutes after she had finished drinking a carbonated beverage. As the two of you are speaking, all of a sudden your friend burps. Loudly. This reaction startles you and embarrasses her, and the two of you have a good laugh. But your brain responds in a particular fashion to this unexpected stimulus. It puts out an electrical signal that in the business is called a *P300*. This signal can be picked up by an EEG, the brain-wave-measuring machine we talked about earlier. A P300 lasts about three hundred milliseconds and hence its name. The same response would happen if you were talking in a room full of people and all of a sudden one of them said your name. Bang, there would go another P300 in your mind. It happens only when you experience a sudden event in a series of partial events you've already paid some attention to, like listening to a person's voice during a phone call and then hearing a belch.

Researchers have used the P300 synthesizing ability of a mostly disconnected brain like Mike's in an attempt to communicate with it. In

Mike's case, the living room had been converted into a futuristic-looking laboratory. Immediately in front of Mike's wheelchair was a board with the letters of the alphabet drawn on it. A light was in the back of the board, flashing on each letter one at a time. On one wall was a computer with a video display and a voice synthesizer attached to the monitor. The voice synthesizer would talk out loud any message printed on the computer screen. Finally, Mike was hooked up to an EEG machine that was also hooked up to the computer. The EEG was responsible for those electrodes springing from his scalp (see fig. 9.5).

All of the equipment and personnel had been choreographed like a well-rehearsed musical. The computer had been programmed to recognize whenever Mike's brain gave off a P300. Mike had been trained to look at the series of letters and pick out one of them in an attempt to spell a message. When the traveling light hit the letter he wanted, he had been trained to get his brain excited. That is, he was trained to get his brain to give off a P300 spike. If Mike succeeded, the computer would recognize the spike and put the letter that corresponded to the spike on the computer screen.

For example, let's say he wanted the letter *E*. He would concentrate on *E*. When the light passed by that letter, he would tell his brain to emit a P300. If he succeeded, the *E* would appear on the screen, and he would be on his way to spelling a message. When the message was completed, the computer would speak the message via the speech synthesizer. Mike's first message was an attempt to greet his mother. And when he succeeded, the computer rang out his words.

Mike learned to give off P300s whenever he wanted to take down a message. Presently, he can do it fairly easily, especially since the letter selection panel has been improved. Mike's brain has been challenged to manipulate a thought pattern not formerly in its power to control. However, his brain activity is the only thing he has left. And the scientists attempted to throw him a technological life line so that Mike could use that activity to communicate.

Controlling the mind like that is amazing, rather reminiscent of those old biofeedback studies of the 1970s. In a very real way, Mike's brain has been able to reach down inside the guts of a computer and use it to communicate with the world. Or perhaps the machinery is acting like a microphone to reach into his mind and let Mike's voice be heard.

These mental gymnastics illustrate a significant aspect about the flexibility of the brain. The brain can continue to rewire and redefine itself long after most other organs have reached their maturity. Or even after they have quit responding to deliberate commands. The point of this section is to illustrate that these kinds of processes are within us, too. The brain's ability to reconfigure itself in response to certain situations is a

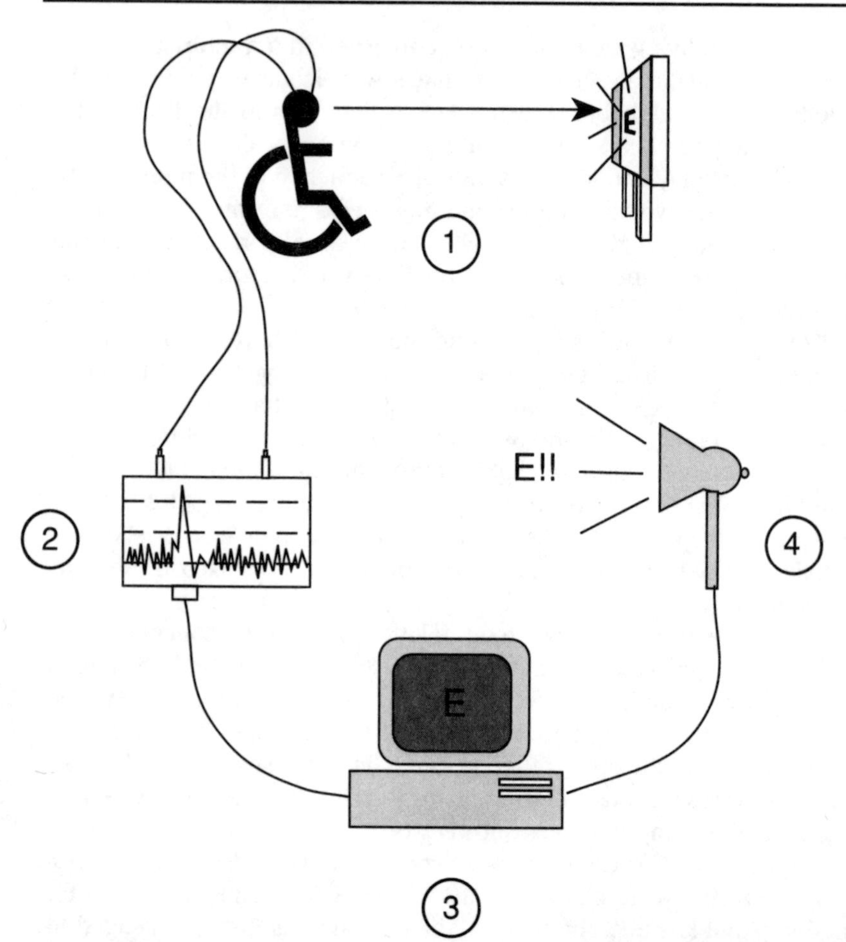

Fig. 9.5. **Signal processing for locked-in minds. Person recognizes a letter on a lighted board (1). Person responds by spiking a P300 wave form, which is read by an EEG (2). A computer recognizes a P300 at the time the letter was selected. The computer puts the letter on the screen (3). Voice synthesizer speaks what is on the computer screen (4).**

normal part of growing up as a child. We even retain parts of this flexibility as adults.

Especially if, for some of us, it is the only thing we have left to offer our world.

A Tale of Two Kitties

Some of the most intensive work describing the plasticity of the brain comes from studies about vision. An interesting fact has been found. Your brain does not learn to see until it has something to look at. Okay, so that runs contrary to what you might normally expect. After all, you look down at a cute little infant, the little guy looks back at you with a vacant expression, and you say, "Hey, he recognized me!" Of course you know it's not true. The eyes of an infant may be able to collect light, and the brain may have developed normally; but that does not mean he registers everything accurately in his visual field. In fact, infant vision starts out between 20/400 and 20/800. There are critical periods of development when visual cues are learned, and if they are not learned at the critical time, they can never be learned afterward.

That there are critical periods of visual development was first shown in a Nobel Prize–winning series of experiments with some unfortunate kittens. Two months after several kittens were born, their eyes were very carefully sewn shut. They spent a period of their little lives absolutely blind. When the sutures were removed from their eyes, the animals remained blind. Permanently. No amount of stimulation could get them to see anything in their visual field.

The experiment was then tried again, but only one eye was sewn shut. Lo and behold, when the sutures were removed from the eye, it was blind. The undisturbed eye had normal vision. A close examination of the blind eyes showed that the eye tissue itself had developed ordinarily; the nerves could respond quite nicely to light.

However, when the brains of these kittens were examined, there was a great alteration in the number and kinds of active neural connections. The brain had become inactivated because it hadn't been used. It was kind of like a muscle; if it wasn't used, the tissue atrophied and became dysfunctional.

These same experiments were tried on adult cats. There was no loss of vision, whether both eyes or a single eye was sewn shut. They tried newborn kittens. To their surprise, there was no loss of vision, either. So then they tried cats of various ages, three days, three months, three years, and so on. They found that during a critical period of time, between the ages of two months and four months, the eyes were susceptible to being permanently blinded. The brain opened up like a flower and was plastic for a period of time, vulnerable to the effects of light. After that time, the brain closed up. It was no longer susceptible to blindness. Thus, the brain was flexible, able to be rewired by conditions affecting it from the outside. But only for what became known as a *critical period of development*. After that time, the brain was no longer responsive to the deprivation.

This experiment was later repeated in monkeys, with very similar overall results. And of course the question was raised, Is this plasticity available in humans as well? Do we have a critical visual period in our postbirth brain development, too? The answer turned out to be yes. The first evidence that this visual plasticity was true in humans came from studying patients who had eye damage—like unilateral cataracts—at birth but were then later surgically corrected.

These patients had such poor eyesight at birth that they could not even see fuzzy outlines of things. As the patients grew older, a new surgical technique was developed that could physically correct the eye defect. At puberty, this surgery was performed in hopes of repairing their eyes so that normal vision could be restored. The result? Nothing. Absolutely no change in vision. As a result of their surgery, their eyes were physically pristine. But it didn't matter. The patients could not see normally after surgery, and they never learned to see normally for the rest of their lives. They did not learn it because they could not learn it. The critical period of development when the brain was going to make visual sense of the world had passed years ago. In fact, if such defects are not corrected by the fourth month after birth, you might as well skip the surgery. These observations and others like it demonstrated the similarity of humans to the cat and monkey experiment.

A Single Viewpoint

This visual learning classroom has since been demonstrated in humans at a more refined level. Its biology is worth talking about because it illustrates the flexibility of our brains. And it has to do with the fact that we have two eyes. Have you ever wondered why you don't see two separate scenes in your mind? After all, you have two eyes. If you shut one of them, the other is completely able to reproduce the entire visual field.

Because you see only one field, you have to fuse your double exposure into a single cohesive image somewhere along the way. You do not have that capability when you are born; you have to learn it. How do you do that? For a period of time after birth, you retain the power to connect and disconnect nerve fibers in the visual center. It's kind of a variation on use it or lose it. When two nerves, one from each eye, see the same visual cue, those nerves become favored. Their connections are allowed to remain active. Nerves that do not see the same thing, that give conflicting images, are not favored. Instead, in an intriguing experience-dependent fashion, they become disconnected.

Thus, your brain is capable of minute refinements and rewirings after it has been formed. The sum total of this activity is that you learn to see one field of view from two images. There is certainly some fine-tuning to this experience-dependent pruning of retinal connections; if this switching were allowed to become rampant, binocularity would be completely disrupted. Nonetheless, permanent biochemical and genetic changes in the neural structure can happen after birth. And they occur because your brain has stuck out its visual tongue and taken a taste of your environment.

The Mind Made in Japan

In terms of vision, thinking of the brain as malleable clay is an odd thing. We don't think of our hair that way; we can't rewire the genetics in our scalp to give us a different hair color. We can't alter the circuitry of our kidney-filtering system if we want to change the frequency of going to the bathroom. But somehow, at least for a while, that is exactly what we do with our brains.

There are examples of brain rewiring other than response to visual cues. Auditory rewiring can occur after birth. The brain molds and restructures itself depending on what it hears when we are infants. This does not mean our brains will look like ornate opera houses if we listen to Puccini or scrambled eggs if we listen to rock 'n' roll. That's what happens to our brains when we let our kids listen to such stuff. Instead, what we hear at certain intervals in our lives affects which part of the brain will perceive certain stimuli.

For example, differences in the way the brain gets wired up depend totally on the language you were exposed to as a child. If you heard a language composed mostly of vowels (like Japanese) when you were a toddler, your brain will perceive emotion on a different side of the brain than if you learned a language composed mostly of consonants (like English). Such plasticity will remain in effect to about the age of nine.

Language learning will also affect the side of the brain you will use to hear musical sounds. If you learned Japanese first, you will hear the sounds of the flute, the violin and most Western orchestral instruments on a different side of the brain than if you learned English first. It even affects which hemisphere is going to be responsible for sexual functioning.

These facts do not have a racial component. Instead, they point to a purely environmental influence on very flexible human gray matter. An American child who learns only Japanese will have the same cerebral dominance pattern as a native Japanese. Conversely, a Japanese child born and raised in the United States will have a dominance pattern indistinguishable from that of American children. Much like our visual system, our brains retain a tremendous ability to be reconfigured depending on the auditory (language) environment we experience as children.

Chances with Wolves

Not only do visual and auditory stimuli effectively change the way the brain acts; there are suggestions that speech can, too. At critical periods of development the brain apparently opens up and slurps in as much language as it can but then, after a period of time, can no longer slurp it up at the same rate. This ability has caused no end of family arguments, especially for Americans preparing to go abroad. The little five-year-old daughter can learn the language and the accent more quickly than her thirty-five-year-old mother can and even more rapidly than her thirty-seven-year-old father can (if he was willing to admit to it).

Numerous pieces of evidence suggest a critical period in language acquisition. Young children whose speech is affected after severe brain damage often completely recover their ability to speak. Adults who suffer the same kinds of injuries often never recover their language skills or recover them only poorly. This evidence, like second language acquisition, suggests that the brain is much more amenable to learning speech at certain stages of development.

An interesting example of language acquisition comes from an event that sounds like it came out of a supermarket tabloid. It is a (rare) documented case of infants raised by animals. In 1920, a wolf mother was discovered in India with two wolf cubs and two human children. No one knows how the kids got there, but their wolflike appearance and animal behavior suggested they had existed in the wild for some time. One child was eight years old and the other about a year and a half. Their discovery ultimately proved their downfall; the younger one died within a year of "captivity," the other about ten years later.

These wild or feral children provide an opportunity to study innate language skills in kids who never learned these capabilities. In the case of the older boy, language skills as we know them were never assimilated; he got only as far as a few words, regardless of repeated attempts to teach him. Such results were consistent with other reports of feral children. It was, in a sense, a natural experiment. Here were humans who had never been taught to speak. If there were critical periods of development for speech that opened and then closed, one might predict that the kids would never learn to speak. And that is exactly what happened.

To be honest, a lot of controversy surrounds the work with feral children. Some controversy centers on the entire idea of critical periods of speech acquisition. These data point to the fact (but do little to prove) that there is a critical period of development for language acquisition. That is, the brain is like a drawbridge that lets certain pieces of information inside for a period of time. When the drawbridge closes, it will not let anything else through, or lets it in at a very reduced rate, for the rest of the human's life.

What is clear is that the brain is plastic and flexible. Whether one is seeing the colors of the world, hearing the sounds of the world, or attempting to verbally address either, the brain can mold and fit to certain desires like jelly. It can run a computer, even if it has only the ability to respond to burps over the telephone.

Difficult Reviews

I once attended a lecture by a biologist who addressed the topic of human cells and human beings. In a paraphrased form, here is part of his lecture:

"When does human life begin? That question can be answered by solving this question: When does human life end? And when does human life end? That question can be solved by asking this question: When does human life begin? The questions are in a circle because, right now, the ideas are in a circle. And a circle may be, in the end, all we can come up with. As a society, we are willing to say that life ends when a human no longer exhibits detectable electrical activity in the brain. If the ending of human life is defined as the absence of brain activity, are we brave enough to define the beginning of human life by the presence of that activity?"

This kind of thinking can stir up a wealth of heated feelings. It certainly did in me at the close of the biologist's lecture. Such thoughts automatically assign worth to the presence of a thinking, functioning human brain. It's not a bad start if the goal is to ask science to define the question for us and then answer it. But can science answer that question? To understand what kind of worth might exist, we have had to define what thinking means and apply that definition to something we can physically measure. It has been the goal of this chapter to tease out some issues involved in trying to do just that. And to show that the effort, mostly because of a lack of information, has fallen flat on its face.

We like to believe that the unique activity separating humans from animals is our ability to think intelligently. Sort of. So a refinement of the idea mentioned above is to detect a very particular type of electrical activity. That is, we should look out for that point when human intelligence swoops into a group of cells and changes their connections. Or gives them a new magnetic field. Or makes them crave calculators on their belts.

The problem is that the biology we are working with is not nearly so cooperative as our desire to simplify it is. Or as easy to convince. Defining the occurrence of human intelligence is about as easy as going into a store with six other people to rent one video. We know that we are generally smarter than animals, especially when it comes to such complex issues as not declaring war on each other or taking care of our environment. These differences allow us to sequester a piece of intellectual turf squarely as our own.

That isn't new information, however. We knew we were smarter than animals back when we thought the earth was a marble table mounted on the backs of four turtles. That doesn't mean that in all these centuries we have gotten a centimeter closer to defining the real difference. We've gotten better at asking certain questions. Or at looking at really intelligent people's brains, acknowledging a difference, and going about our business as if we had learned something.

Facts to Grind

Yet we have kept trying. The hope of pigeonholing intelligence into specific categories was that we could divide and thus understand intelligence in terms of territories. The hope of formulating something like IQ was to perhaps quantify and compare those territories with each other. It hasn't worked out. We find ourselves in the intellectual equivalent of a spin cycle. First there were 2 categories of cognitive abilities, and then there were 120. People were smart because they performed well on certain tests; they performed well on certain tests; and therefore, they were smart. That reasoning is inadequate to explain anything within its own framework, let alone try to apply to solid biochemical mechanisms. It doesn't mean that these weren't and aren't noble efforts to get at an extremely complex and vague process. But at this moment, it's like trying to predict the weather by describing the imaginary shapes we see in the clouds.

Physical attempts to define the occurrence of intelligence have not clarified much either. When we break into the private life of a human fetus, we can examine cells that move from point A to point B. We don't know if the movements hide a budding Stokowski or portend an undeveloped Elvis. We don't know if the presence of this chemical or the absence of that connection has a great deal to do with Junior's ability to write poetry or grasp quantum physics. We have not yet found a way to ask a fetus if it is talented, and we are even a longer way from asking if it is smart.

The neurobiology of intelligence isn't all smoke, however. We can reach definite conclusions by examining certain neural growing pains. We know, for example, that when a particular structure in the brain is present, a particular intellectual function may occur. It's like saying that when a bicycle is present in the driveway, it's possible to run over it. That doesn't mean it will be run over. That doesn't mean it won't be run over. It's just that the opportunity presents itself. In the same way, we don't know if the presence of certain human neural structures blares out to the embryologist that intelligence exists. The researcher probably wouldn't identify intelligence if it reached out and grabbed him by the test tubes anyway. It will not mean, however, that a certain structure does not possess intelligence. And it will not mean that the structure has possessed it all along, either. At this point, we just don't know.

The Missing Think

We can determine some things if a particular structure is absent. We can say that if a particular structure does not exist, its function cannot exist, either. For example, if airplanes did not exist, it would not be possible to ride in them. In the same way, this definition puts certain capabilities into a framework. We know that a three-week-old embryo cannot count to three because there isn't an adult nerve cell in its body. A twenty-two-week-old fetus cannot read Descartes because the nerves that allow him to gaze at books have not been fully developed. A two-minute-old newborn cannot fuse both fields of view into a single cohesive image. And an adult will never learn to deliberately spike a P300 unless he or she is trained.

There is a problem in defining intelligence by such examination. These issues assume that to some extent, human intelligence is a static concept that can be localized to a set of nerves whenever a particular structure is visualized. That is probably not an accurate assumption because the brain is not a settled-down, learn-everything-once-and-then-quit kind of organ. It is an extremely plastic, flexible organ. It continually rearranges its architecture and then, at certain critical periods, changes its capabilities. There are critical periods of development involving sight, sound, and speech when this dynamic brain surfaces with amazing capabilities and after a period of time—and for unknown reasons—plunges back to the genetic depths never to rise again. Even when it is totally isolated from its normal hookups, as in Mike's case, it is capable of learning new tricks. That is Pliability with a capital *P*. An all-encompassing definition of human intelligence will have to include the fact that the brain is capable of changing and redefining its talents. It can happen long after birth.

No Lurking Definitions

So what does all this have to do with the acquisition of human intelligence? Probably nothing. And that may be the point. At this stage in our technology, we comprehend boundaries, not answers. This is troubling because we are asking questions like, What is the castle made of? And we are getting answers like, It's two and a half miles down the road, then turn right. Both ideas are relevant to the castle, but one involves content and the other involves location. At this stage in our technology, defining when human beings get their once and future unique cognitive abilities is quite impossible.

And that is the frustrating thing. So much is riding on our answer. Many would like to believe that science can tell us when human intelligence arrives at the doorstep of a uterus and begs an audience with the developing baby. The hidden hope is that this will also tell us when a human tissue becomes a human being. As we have seen, science is having a hard time coming up with a definition of human intelligence. Let alone explaining why one group of cells decides to nibble at a garbage can I am aware of and why another group tries to jolt the first group out of existence.

Chapter 10

Knot Within Reason

We knew a couple with the sweetest little girl you ever saw. She had big blue eyes that absolutely begged you to spoil her rotten, a beckoning no rational adult could—or did—ever refuse. She was also extremely inquisitive, full of the curiosity and vitality so characteristic of bright, well-loved five-year-olds who get a lot of attention. My wife and I had a clear demonstration of that curiosity one day when her parents invited us to a picnic in the mountains.

While we were hiking up to a meadow, the little girl ran across some flowers open in full bloom. She promptly asked her daddy what was inside flowers that made them open so wide. Her daddy told her that the plant was alive and the same life that made little girls laugh and skip rope also made little flowers blossom. She then asked her daddy what the difference was between little flowers and little girls. And that is when her dad gave a sheepish grin and glanced at me. I bravely avoided the subject by turning my head away, using my best preoccupied objective scientist attitude, pretending not to hear the conversation.

But I did hear the conversation. And it got me to thinking about her question. It got me to thinking long and hard about who we are from a scientific point of view, a social point of view, a religious point of view; in many ways, the book before you is the result of those ruminations. I have at all times wanted to avoid the psychological aspects of this identity crisis, a subject already the preoccupation of an increasingly unnecessary number of books. What we consist of biologically, deep beneath our flesh, buried in the watery grottoes of our genetics, has been given much less rigorous treatment. By that I do not mean anatomical descriptions. There are certainly enough anatomy and physiology textbooks to rival those on the personal identity shelf. These must necessarily explain only biochemical aspects of tissues, as marvelous as those aspects may be. In the end, we have not touched on our inward identity any more than we have described how a certain pastry tastes by providing a written description of its shape.

The purpose of this chapter is to ask if science is capable of telling us about such a taste. We have been exploring some of the processes involved in defining life and death and how humanity is baked onto each of them. Now I wish to draw the various ideas together and give research a report card, assessing science's ability to discriminate among these processes. To do that, we must first ask if research is able to draw us any closer to a cohesive definition of life and death. Then we must ask if it can do the same in defining the difference between human beings and human tissues. We may get a better handle on the ability of research to give us insight into our own lives. It is the only thing I could think of, short of shrugging my shoulders, in seeking the differences between a flower and a child.

Fears of Influence

Science is a lot like my friend's precocious, strong-willed daughter. You can ask science a question, but there is absolutely no guarantee that you will get a straight answer. Or that you will get an answer you are looking for. You may get an answer that reveals holes in your question, which is about as gratifying as a mosquito bite. I am constantly amazed that a discipline needing to exert such control over its environment is, in many ways, so completely uncontrollable.

Many people misunderstand this scientific love/hate relationship with control. The confusion tends to center on two areas. One misunderstanding involves a misperception of the power of the researcher. A second misunderstanding involves a misperception of the power of the research. These misunderstandings are sad; the tendency to blur the human with what he or she has discovered in many ways distorts the beauty of both. And in the end, ascribes to each more power than they intrinsically possess.

Blunder Enlightening Storms

Let's first concentrate on the researcher. I believe a subtle religion has come into being in the twentieth century. I call it the First Church of the Lab Bench. It has a priesthood, consisting mostly of scientists, who celebrate a communion with truth by chipping away at the unknown. Our attitude toward this priesthood is often one of awe. We behold satellite transmissions and miracle drugs and weird genetics and think that because scientists can do so many wonderful things, they can do *all* wonderful things. These discoveries seem to have given this priesthood inherent magical attributes that, in turn, have given it social power.

Although that is a nice ego boost on days when grants lose their funding, it is also a lie. Researchers do not automatically possess the majesty of the things they examine just because they study them. Blurring the research with the researcher is a lot like hearing a beautiful symphony on the radio and then complimenting the stereo speakers on their compositional skills. The wrong entity gets the credit. The accolades given to researchers spring from their ability to see; they can be acknowledged only as sources of marvelous insight into what they study or can be affirmed for their in-depth knowledge of the literature or their ability to successfully socialize with the people who hold the money. In the end, Nobel Prizes are awarded because the investigators have good scientific eyesight. Period. When we transform the scientists into their research, we erase the boundary between subjective personalities and objective creations.

Which is, of course, the twentieth-century version of worshiping the golden calf.

This distinction can be amusing. You would not believe the significant results that come about by sheer accident. By accident, I do not mean to imply disarray. As scientists, we must follow certain very tightly controlled rules to figure out something. But because we do not know what we will find, we have absolutely no rule to follow as our results unfold. Most of us are doing the intellectual equivalent of bird-watching, wandering around in the woods trying to adjust our field glasses when boom, we crash into a branch of truth. Many—I am tempted to say most—significant findings have been stumbled upon in such a fashion.

Investigators who are particularly successful are not necessarily particularly talented. Some are particularly lucky. That truth does not stop certain segments of the public from being in awe of what we come up with in the laboratory, however. Considering how much gets muddled, quite frankly, so are we.

Intended for Measure

This brings me to the second point concerning the power of research science. As just stated, it is a mistake to project the power of research onto the personality of the researcher. An even greater error is made, however, when the power of research is projected onto areas of our lives that science has no power—and no business—measuring. For example, many people believe that this church can somehow save us out of all the dilemmas we create for ourselves. It is almost as if the scientific successes of this century have intoxicated our intellects, filling them with incredible hallucinations about the present and wild hopes for the future. Even worse, these fantasies take on an almost satirical complexion when science is used to quantify areas of the human experience it is totally incapable of measuring. What is missing is a realization that science can measure only certain things. And what are those certain things? Here is a statement that should be written on the top of the door frame of every academic department of every college in the entire planet. It is an axiom:

SCIENCE IS CAPABLE OF MEASURING ONLY PHYSICAL THINGS

That's right. Researchers are capable of measuring only physical, you're-matter-or-you're-energy or you're-both-but-that's-it phenomena. Science is absolutely incapable of measuring objects or events that are not physical. It is equally incapable of measuring objects or events that are partially physical. It is incapable of measuring objects or events that are mostly physical. If it's not totally physical, science can't measure it. If it is totally physical, science can measure it.

Here is a partial list of things science is capable of measuring:

- The speed of light in a vacuum
- The number of cells in a two-week-old human embryo
- The wavelengths of light emanating from a distant star
- The average length of a hot dog in the deli section of Dissmore's Grocery Store in Pullman, Washington
- The weight of Roseanne Barr after she finishes eating three bacon cheeseburgers at a local fast-food restaurant

Here is a partial list of things science is *not* capable of measuring:

- How beautiful a sunset is
- How ugly a sunset is
- The amount of musical inspiration in Stravinsky's *Le Sacre du Printemps*
- The sense of joy experienced by a woman at the birth of her newborn baby
- The sense of loss experienced by a woman after her first miscarriage

Whether we are shaking with laughter or drowning in tears, science is incapable of measuring many things in the human experience. This lack of calibration does not mean these experiences don't exist. It simply means that science can't measure them. Such investigative inadequacy demonstrates that the power of the First Church of the Lab Bench is limited. In the end, we send humans to the moon and clone genes from heart cells only because we have developed fancy yardsticks. But we cannot calculate the amount of thrust we will need to send feelings of grief to Mars. Or clone the joy genes from myocardial tissue. We cannot use investigative research to answer these questions because grief and joy are not physically quantifiable events. They may arise from physical beings; they may exert a physical effect on the being from which they derive. It doesn't matter. They lie outside the grasp of the objective. It makes about as much sense to use science to describe these experiences as it does to use a yardstick to measure how much you weigh.

This blurring is in some fashion understandable. It is hard to see that science, especially as it is practiced today, is only a human endeavor. But it is a human endeavor, replete with all the strengths and all the failures of any human enterprise. With equal enthusiasm, scientists have created technologies that either nibble holes in our ozone or increase crop yields in Third World countries. It can be argued that many of our present-day human ills—from pollution to nuclear weapons—came about as a direct result of scientific effort. To ascribe to science a certain amount of wonder is to blind ourselves to its failures even as we prostrate ourselves to its successes. Science as an intellectual discipline makes a wonderful professor, but as a spiritual religion, it makes a lousy god.

Coping with Life's Ups and Downs

How can a discussion about the limits of science help us to discern the difference between human life and human tissue? At first glance, science may appear to be able to answer the question about tissue and life quite easily. Human beings are physical entities with defined characteristics. When we are not fighting wars or arguing about the federal budget, these characteristics generally separate us from other animals. Human tissues are equally physical entities, also with defined characteristics. We should be able to assess both sets of characteristics, find which are absent in one and present in the other, and come up with a working definition based on those differences. Right?

Yeah, right. The problem is that biological entities, while they may be physical entities, are extremely complex physical entities. And even more so, human being physical entities are capable of creating and comprehending ideas that are not physical at all. To answer questions about human life, we have to ask questions about life in general. That means going to the dictionary of biological research and looking up the experimental results that would lead to definitions. As stated in the first chapter of this book, such steps are broken down as follows:

1. What is biological life?
2. What is biological death?

After obtaining working definitions of these two ideas, we would be ready to superimpose upon them questions regarding our own species. These questions can be broken down as follows:

3. What is human life?
4. What is human death?

We then need only assess how well science addresses these questions to know if it is capable of answering them. It's a lot like asking if someone is home by ringing the doorbell. If someone answers the door, the person also answers the question. If the question about the difference between human life and human tissue is scientific in nature, science can answer it. But if there's no one home in the scientific household, if an answer doesn't come when we ask for the difference, we are knocking on the wrong door. And so the nature of this book has been to rap as hard as we could on the front porch of investigative research and listen for any vague stirrings in the house. The question we must ask now is, Does anyone come to the door?

Molecular Coordination

In previous chapters, our first series of knocks consisted of examining the concept of biological life. We found that science was fully capable of describing the attributes of living things, which for reasons of job security is reassuring to most biologists. Some tissues were extremely hardy, performing nearly miraculous feats of endurance, stoking the fires of existence, ensuring our survival. At the same time, science described tissues and biochemistries of extreme delicacy. These tissues had survival requirements that bordered less on the miraculous and more on the ridiculous.

Science found that all of these systems, whether hardy or fragile, worked together in a coordinated fashion. Some of the coordination involved specialized cells that interacted only with other equally specialized cells under very finicky conditions. Some of the coordination involved cooperation between giant banks of cells whose molecular task masters could be other cells or very particular biochemicals. The picture that emerged was an infinitely complex ballet of tissues and energies and functions, all of which could be summarized as a giant collection of molecules. If there was a particular organization of these molecules, the collection could function in particular ways. The sum total, under specific conditions, could be called life.

Putting the Part Before the Horse

If we could leave biological life at that level of descriptive simplicity, we could go about our merry way and leave philosophical discussions for those branches of academia that don't generate grant money. Unfortunately, most organisms that we call living undergo reactions based on constituent parts. All organisms are composed of organs, which are composed of cells, which are controlled by nuclei, which are described by chromosomes, which are made up of genes. And the great problem is that science can also examine these constituent parts and find them to be functional even when they are separated from their natural living environment. Under those conditions, science's ability to describe life begins to break down. Examining life scientifically changes the nature of the question.

For example, science does not distinguish between the life in a human patient and the life in one of his or her genes. This lack of discrimination was made clear in an experience I had some time ago. I assisted a group of researchers in cloning a gene from a patient's blood vessel. A sixty-year-old white male came in code to the emergency room. He had died from a massive head injury and was an organ donor. After the release forms were signed, his tissues became available for scientific study. A resident I was training went in to harvest a section of the aorta of the patient. As is always true with such bodies, the resident beheld an eerie sight. There was a man, perfectly healthy in every way except for critical brain functions—a body fully alive and, at the same time, a person fully dead. The patient was not called a human being but was termed a heartbeating cadaver.

The aorta, once removed from the body, was packed in ice and quickly brought into the laboratory. The cells of interest were gently removed from the vessel and placed into tiny round plastic dishes. The remainder of the aorta was put into a plastic bag to be discarded at the end of the day. The cells were fed with nutrients and antibacterial reagents and then put in an incubator. The next day as we examined those living cells under the microscope, I felt those same eerie feelings. The cells were fully alive and reproducing in my petri dish, but the host and the blood vessel had long since expired.

I remember wondering how science could define life when an organism was no longer living, yet possessed parts that could still be called alive. At that moment, the only difference between the human in the grave and the cells in the petri dish was their functional organization. In attempting to understand exactly what was still alive, I was forced to ask exactly what had died.

The next step after allowing the cells to grow in culture was isolating the genes locked deep inside their nucleis. We did it in the most impersonal way. The cells were scraped off the petri dish with what amounts to a spatula and dumped unceremoniously in what amounts to a blender. They were then ground up with a mixture of chemicals. The procedure utterly destroyed the life of the cells. It also separated the genes from the rest of the molecules in the cells.

After several manipulations, the genes were placed in viruses where they could be more readily handled. Then it was easy to isolate a particular gene, which was eventually placed into a bacterium for permanent storage. The gene was fully functional inside the microorganism. RNA was made from the gene, and the protein was made from that RNA.

The protein, when isolated, performed functions it had previously done only in the man. And once again, that eerie feeling came over me. The gene, a copy of which once existed inside a human being, was alone inside a bacterial tomb. And even so, it exhibited life, or at least some kind of function. At some tiny level, the gene was repeating a process it had learned from a human embryo that existed twenty-six years before I was born.

What is the point of this description? The existence of functional component parts in the face of dysfunctional whole organisms brings forth a very important idea. It is not as easy to understand the existence of life when it is placed in the test tubes of the laboratory as it is to understand the existence of life when it is found in the grandstands of a football stadium. Science has a tendency to look at the viability of tissues and genes from dead organisms and say, "What exactly do you mean by life?"

You can ask the same question in this example. What is truly alive in such a patient? His aorta? His cells thriving in culture? The genes? Is life transferred like a football, handed down from organism to molecule as we whittle away at component parts? Studying the biochemistries of living things invariably forces us to consider not what is objectively alive but what is objectively functional. And when it comes to obtaining a definition of life, we must either change the nature of our question or change the nature of our ideas.

Gull Friends

I had another opportunity to experience those eerie feelings about life and death. I was walking along the beach with a close friend early one morning. It was a morning that positively sang with avian motion, sea gulls filling the air like a giant cathedral choir. They so captured our view that we nearly tripped over a rather nasty pile of feathers and decaying flesh lying in the sand beneath our feet. There were two dead sea gulls, one looking as if it had just died and the other like it had been dead for some time. Seeing dead sea gulls on a beach, especially with the number of birds in the vicinity, is certainly not uncommon. But the contrast between those in the air and those in the sand was quite striking. So striking that for the rest of the morning, my friend and I engaged in a conversation about the meaning of biological life and, most important, the facts surrounding its cessation. The fruit of that conversation I relate here.

Elective Biology

As stated earlier, biological death is not a mandatory biological requirement for all organisms. For most living things, death as we commonly understand it is an option. The mode of reproduction often dictates the ultimate fate of the organism. If an organism exercises the reproductive math trick of multiplying by dividing, death is elective. Only for organisms that reproduce sexually, like sea gulls, does death become compulsory. Death is not a monolithic process, even from a nonscientific point of view. That is even true, in a certain way, for mammals of which our species is a part. We may ask death not only what has happened to its sting but also what has happened to its consistency.

More than consistency is at stake, however. As the chromosomes that govern biological processes are better understood, research science is beginning to question death as a valid biological description of anything. The world of the gene, as we have previously discussed, is forcing us to consider the presence or absence of functional organization as a better description of the loss of vitality; death may be an arbitrary description we have made to color certain changes in molecular organization. Actually, we might have predicted that research would have a hard time with a definition of death. If science doesn't say much about the presence of biological life, it is probably not going to say a lot about its absence, either.

A Wrinkle in Slime

An examination of creatures that can exert tremendous physical changes in their structural organization spotlights some of this trouble. In chapter 3, we discussed the little slime mold, the creature that can exist as hundreds of free agents or as a collective giant organism. You recall that upon starvation, myriads of tiny slime mold amoebae gather together and aggregately form a single large slug. When the food supply is restored, they have the ability to disengage and return to their individual amoebalike states. This organism is an example of an animal in a perpetual state of transition, a transition that varies as a function of food supply. The giant slug does not necessarily die because all the amoebae crawl off each other and go their separate ways. Nor do hundreds of amoebae die because they have gathered themselves together to form a large slug. *Dying* is not a relevant term here. The key word, rather, is *change*.

You might argue that the life cycle of a slime mold could hardly be called the best example of death. Organisms roll over on their backs, quit breathing, and begin the decay process. Even slime molds are capable of dying if you pour acid all over them. And they will die whether you pour acid on them at the amoeba stage or at the slug stage. How could death be defined simply as a transition in these or any other animals? Why would scientific eyes leave out the very state so obvious to human eyes?

To answer that very legitimate question, let's refer to the sea gull example. All of us, scientists or not, can watch a live sea gull overhead and a dead one on the ground with two points of view. First, we can be qualitatively aware that great differences exist between them; we observe that one can move and the other cannot. However, a second level of biological description goes beyond the visual inspection of the dead bird. Though appearing lifeless, it possesses lots of biological processes, all deeply at work in its body.

The sea gull is decaying, primarily due to the feasting of microorganisms. Thus, a great deal of life is left in the sea gull after it has died. This life is now packaged in tiny bacteria and performs different functions. They munch so freely because the flesh no longer has an immune system to keep them at bay. There has been a change in the organizational complexity of the molecules in the sea gull—a change that allows the now quite smelly activities of decay to occur.

This decay permits us to eventually detect gross changes in the overall shape of the sea gull. Nothing has been lost, at least from a strict scientific point of view. All the atoms, electrons, and attendant energies are around. If you did your bookkeeping correctly, you would be able to account for all of them. When microorganisms are dining on the expired bird, they are taking sea gull molecules that existed in one configuration and turning them into another configuration. This configuration may be more beneficial to the energy requirements of the bacteria. In the end, the only difference between the bird and the microorganism is the organization of molecules. When science looks at a dead sea gull on the beach, it does not see a cessation of biological function. Rather, it sees a change in biological function. And that is the point. What most of us have defined as death, science has defined as transition.

Better Life Through Chemistry

This odd view of the world gets weirder when we consider the genes governing these states. Science tells us that life/death transitions are the result of the presence or the absence of tiny biochemical reactions. An example of this reductionism is the discovery of what may be the smallest unit of life. In chapter 3 we discussed an extraordinary organism, the plant-menacing viroid. The organism consists of a single gene. And this single gene has one genetic instruction: make a copy of itself. That's it. It has no other purpose in life. The copying mechanism is a single set of biochemical reactions.

Let's explore that idea. If the viroid is in the proper cellular environment, the biochemical reaction will take place, and the gene will be copied. If the viroid is not in its proper cellular environment, the biochemical reaction will not take place, and the gene will not be copied. There is no mystical life force in this viroid. There is no spark of life. It is a chemical reaction. With the proper reagents, you can make such an organism in the laboratory. It will not be any different from the one found in nature.

You can look at things more complex than a viroid, and you will still find genes, governed by the same biochemical reactions. You can look at viruses, with a number of genes working in concert. Make it more complex and you get a bacterium. Or a cell. Or a group of cells. Put the cells together with their genes working in a particular fashion and you will get an organism, a bird or a human. All of these organisms are governed by little genes, blinking on and off in their cells.

Gene activation bears directly on the ability of science to define life and death in complex organisms. The only difference is that now there are many genes, which means there are combinations of on/off states. These on/off switches are not a mystical life force that we have no ability to measure. They are a series of marvelously predictable and artificially reproducible biochemical reactions. That's why we can take a mouse gene and stick it into a tobacco plant, and the mouse gene will be functional.

Let me summarize. Biological life is, at one level, not anything mystical but the sum of those reactions. And death is ultimately a simple absence or change in those reactions. Biological activity is thus a continuum of genes in various stages of organization and functional complexity. This takes away none of the majesty of a sea gull in the air or the sorrow of one in the sand. Instead, it provides a clearer way to understand them.

A Calf Infection

I knew of a top-flight molecular biologist whose father died quite suddenly. His grief over the loss of his dad was strong and intense; I am quite sure he will grieve in some fashion for the rest of his life. Through his many tears, this researcher experienced, as every human experiences, the reality of life and death on this planet. No power of science, no understanding of genetic processes or chemical reactions, could take the place of a loving father and friend. And so this researcher, this gifted scientist, would often close the door to his office in the lab and quietly sob his afternoon away.

Watching someone go through such a loss is always a profound thing.

Nonetheless, because of the way science views life-and-death experiences, there is a tremendous temptation to say that scientists believe these processes do not exist. That the whole world is nothing more than cold impersonal molecules existing in the netherworld of alternate configurations. I know that I have been asked numerous times in social situations if my training affects the way I personally view life-and-death processes. Or at least affects the way I experience them. And the answer to both questions is no.

When scientists—or anybody else for that matter—experience joys or losses, they are not experienced in a vacuum. My friend and colleague grieved not out of his science but out of his heart. And that is the interesting thing. In poetic terms, the heart is certainly capable of understanding the difference between life and death even if the mind is not.

Most of the confusion about life and death centers on the misinterpretation of the power of science. Because science cannot measure the substance of life and death does not mean these processes don't exist. It simply means that science cannot measure them. The problem is not in the essence of these processes but in the tools we use to ascertain their existence. Life and death, though touching physical biological things, are not quantifiable concepts. They are qualitative assessments, generated by organisms with enough intellectual muscle to understand those qualities. Since science can measure only physical quantities, it cannot measure the life and death in an organism. Just like it cannot measure the depth of my friend's grief over the loss of his dad. When science is asked to define a qualitative judgment, it stops making sense; research cannot be asked to describe something it is not capable of measuring.

In the end, it is proper to say that there are some real differences between sea gulls soaring in a blue sky and sea gulls decaying on the beach. It is just as proper to say that science is blind to some of those differences. If the only confusion is a misunderstanding of the power of the scientific twentieth-century idol, we are on predictable ground. However, when we mistake the investigation for the investigated, the first thing we usually do is to poke ourselves in the eye. That, too, is familiar territory. Since the time of Moses, people who worship golden calves have always been blind to reality.

Dark Side of the Tomb

So, investigative research doesn't seem to communicate much about life and death. Science can appear so smart in some areas of our lives and so stupid in others. Like automatic teller machines. Science can study biological life for hundreds of years and tell us many things about it. Everything except what it is. Science can study organisms whose life appears to have left. It can understand this lack of functioning down to individual molecules—and have absolutely no idea what exited. We must understand the wisdom of this stunted genius, in terms of its power and its limitations.

With this warning in mind, we are now ready to ask if research can answer the main question of this book: Is science a tool that can assess the differences between human life and human tissues? In previous chapters, we have discussed characteristics of human beingness. We were attempting to come up with physical standards—a set of criteria actually —for defining the presence of sentient humanity. The issues that have been mentioned, and have been argued publicly, are as follows:

- The physical appearance of the tissue: Is it a human being if it looks like one?
- Human uniqueness: Is it a human being when a uniquely human trait becomes established in a complex group of cells?
- Genetic uniqueness: Is it a human being if it has all the genetic information and environmental support to become one?

We will examine these ideas one at a time. Our task is to see if the criteria, at their heart, contain ideas measurable by scientific methodology. If they are, we can truly use science to discern when human sentience is conferred on human cells. But if they are not, this issue will go the way of biological life and death mentioned in the previous chapters. That is, it will be rendered as scientifically irrelevant even as its importance is infinitely profound.

Familiarity Breeds Contentment

I have talked with many people who say that the humanity in a fertilized egg at the four-cell blob stage is quite a bit different from a living, breathing baby. The argument is that because of the dramatic difference in physical appearance, one is quite obviously a human tissue, and the other is quite obviously a human being. This, of course, implies a difference in their value. We throw tissue away as easily as we cut our fingernails, but it is considered murder to do that with a human being. By physical appearance alone, groups of cells are assigned net worth. Since the cells start out with no worth and end up with an infinite amount, they have obviously acquired it somewhere along the way. For those who hold to the idea of physical appearance, the question is, What was acquired that was powerful enough to create such a drastic change in value?

We first have to establish the validity and the content of physical appearance before we can go after specific sets of traits. What does it mean to look like a human being anyway? There are people who say that a four-week-old human embryo looks like a salamander. The operative word here is *looks.* Some features, depending on what you are examining, absolutely distinguish a human fetus from a salamander. To the genetic engineer, the genes don't look like a salamander at all; they look absolutely human. To an embryologist, a human fetus exhibits only superficial similarities to an amphibian. Only the inexperienced eye makes the comparison to this salamander. So the best question is, When does it look like what I am familiar with? And the answer is that it depends on what you are familiar with.

A Platter of Opinion

This observation—the dependence of evaluation on experience—yields a clue concerning science's ability to discern humanity based on appearance. How familiar do tissues have to become in order for them to be human? An embryo at five weeks might look like a perfectly familiar human to a researcher and not look like anything familiar to a plumber. Do we have two different humanities as a function of the background of the observers? If a baby is deformed enough, even if it is capable of a productive future, is it to be denied human worth solely because of an unfortunate appearance? How about a baby that looks normal in almost every way except that it is born without a brain? Is it to be given human worth, placed into the custody of its parents, and given voting rights when it turns eighteen?

The answer to the question about when something looks like a human is going to mean vastly different things to different people. The rejoinder is not easily quantifiable because the question does not address tangible quantities. The answer is derived from a series of subjective opinions based primarily on individual experiences. Because the question cannot address physical quantities and the answer can consist only of subjective opinions, we have booted ourselves out of the scientific ballpark. Remember that science can address only physical processes. And a subjective opinion is not a physical process. Research can't tell us when a human being exists based solely on appearance. And no amount of experimentation will ever let it.

A Head of the Game

Some persons believe that appearance is not a strong enough criterion alone for determining the presence or absence of protectable humanity. Instead, physical appearance must be accompanied by a uniquely human trait. Otherwise, a corpse is as sentient as a crying baby, a mannequin as valuable as a human. The trait that has been publicly argued as the criterion for separating live humans from animals and cadavers is our rather pronounced ability to think. When a group of cells carry a thought, they also carry a value. Or a business card, telling us that we are addressing a human being. Science has to tell us when a collection of human tissues are capable of thinking in order to also tell us when they are human.

Is this any more valid a measure of the presence of protectable humanity than appearance? As mentioned in chapter 8, there are problems with these ideas. Identifying a uniquely human thought can be terribly difficult. Especially when one considers the state of today's educational standards. Monkeys and gorillas demonstrate many types of thinking, including some of our problem-solving skills. The genes that govern those electrical processes are, in many instances, nearly identical to our own. Moreover, many kinds of human thoughts, including involuntary ones, don't necessarily circumscribe sentient humanness. But they are uniquely human, even if they are generally confusing. For example, brain-dead people can possess electrical activity that keeps certain functions alive. These are human thoughts. Anencephalic children, babies born without a complete brain, often possess electrical activity in the parts of the brain still present. These are human thoughts, too. But whether the bodies that possess such thoughts also possess sentient human life is open to discussion.

It can be argued that we are not looking at the "right" thoughts. Certainly, monkeys could no more compose a ballet than they could grow wings. Perhaps for a group of human cells to possess sentience, advanced thoughts, thoughts capable of generating *The Firebird Suite,* or designing jet airplanes, must be present. Once the thoughts that separate us from animals are in place in our cells, those cells must be equipped with human sentience. Right?

Well, not really. Considering advanced thoughts as a criterion produces a problem. Humans are not automatically conferred with the ability to solve complex problems because they have newly constructed a brain. If we call a group of cells a human being only when they are capable of certain unique problem-solving skills, we would be free to perform hideous genetic experiments on half the undergraduate students I have lectured. Many unique brain connections that will allow complex problem solving are not formed weeks—perhaps years—after a baby is born. Clearly, we cannot talk about the arrival of sophisticated human thought as the arrival of sentient humanity. We are forced instead to talk about the potential for complex thought, whether we are comparing a sixth grader with a graduate student, or a human conceptus with a toddler. The potential for human thought, heralded by the arrival of the neural substrate, is perhaps the process with which we can judge the presence of human life.

Philosophical Substance Abuse

These potential arguments make it difficult for science to come up with a definition of humanity based on unique thinking. Science would have no problem discussing the development of neural tissues. Or even looking at the presence of brain waves. But it might have a big problem drawing the line between a stage that has enough potential and one that does not have enough potential.

At what point is the promise, rather than the substance, strong enough to carry the weight of sentience? The possibilities stretch back a lot farther than brain development in the first or second trimester. The secrets are locked in the genes. And the genes are in sperm and eggs. The beginning of the potential for human thought can just as easily exist in the construction of the genetic information as it is in some half-finished brain of a three-month-old fetus. It all depends on your point of view, not on an objective standard.

And that's the idea. Since points of view are not scientifically quantifiable entities, we cannot use investigative research to assess their validity. Whether the undeniable potential of an event is just as compelling as the actual event is not something that can be experimentally determined. And it is true in this biological question. Science can't judge whether the presence of functional genes that will eventually code for unique human processes are as real as the actual processes themselves. And since these issues are a matter of subjective opinion, they are also a matter for some tool other than science to decide.

A Line in the Strand

Some people say that once a sperm and an egg have joined together, this unique combination has carved a human being out of what was only human tissue. As far as they are concerned, when science tells us that the nuclei have fused, science has also told us that a human being has formed. The distance between a human being and a human tissue is the distance between an enthusiastic sperm and a willing egg. And after this distance has been closed, the genes placed together are of such extraordinary power that a human being is formed.

Is that true? And if it is true, was science the one who told us? Is this collision of cells the unique event that makes individual human beings out of previously unjoined discardables? Can experimental research inform us about the onset of a unique human life? The questions here have to do with the fate of the fertilized egg, the absolute uniqueness of the joining, and science's ability to say anything about humanness once the joining has occurred.

No Unique Beginnings

Once an egg has been fertilized by a sperm, we will not know lots of things about its uniqueness. For example, we won't know for a couple of weeks if this egg will be a single unique human being. There is a certain chance that it will become two human beings. Or triplets, in which case we really had multiple uniqueness at the beginning of conception. Second, we can't predict the destiny of the fertilized egg during the course of the pregnancy. Many pregnancies, as many as one-half in some studies, will end in a spontaneous abortion. For any given fertilization, we don't really know if a football star or a miscarriage is created. Moreover, a certain percentage of these egg/sperm collisions will end up deformed; one in two thousand will be anencephalic. Many of these mutations will not survive the pregnancy or, after delivery, survive outside the womb. It can be argued that some of them never lived because their tissues were never organized enough to be anything other than a large mass of cells. Like a tumor. We may have the beginnings of a human life, the beginnings of a miscarriage, or the beginnings of a deformed, partially differentiated mass of cells in any given conception. From a scientific point of view, we can say very little about the overarching presence or absence of a unique human being simply because egg and sperm have joined together.

Chip Off the Old Clock

To be fair, it can be argued that fate is hardly the criterion for determining the absolute presence of a human being in tissue. That is certainly true, especially if one operationally defines the presence of a human being by the creation of a fertile time clock. (By time clock, I mean this: a sperm and an egg joined in the womb have a nine-month clock that is wound up and released at conception. A sperm and an egg separated from each other in a womb do not have that clock and can never create a human being.) The problem is that a fertilized egg has a finite number of options it can follow, only one of which will result in the creation of a human baby. From a scientific point of view, a two-celled embryo does not perform the same biochemistries that a 2-billion-celled infant does. It is thus not the same thing as a 2-billion-celled infant. A fertilized egg has the potential to create a newborn baby. But two minutes after fertilization, it is not a baby in the same way that one who has just been born is a baby.

But are they human beings? One can argue that a fertilized egg isn't a baby, but one can also argue that a fertilized egg isn't an adult. That doesn't mean that an egg and an adult are inhuman because they are unlike each other. One can say that each has different levels of molecular organization.

We can certainly postulate that a fertilized egg is a human being. Some persons would say that the initiation of the clock is a good enough reason to call any fertilized human egg a human being. The problem is that science cannot evaluate what a good enough reason is. We could just as easily say that the presence of this clock is not a good enough reason to call a fertilized egg a human being. Science would have just as hard a time with this statement. Why? Because a good enough reason is a value judgment, not a scientifically verifiable idea.

Science doesn't say anything about it because science can describe only interactions between molecules and how fast light travels in a vacuum. Because a good enough reason is actually a value judgment pronounced over a certain set of facts, the idea is automatically kicked out of the realm of science. It might appear coldhearted to say it, but science cannot tell you if something is of value. Therefore it cannot tell you if something should be saved. That doesn't mean the fertilized egg is an impersonal mixture of carbon and hydrogen. But that doesn't mean it is a priceless and miraculous human being, either. Science isn't the tool that can tell us one way or another.

A Down to Worth Idea

What is the problem here? Why does science throw curveballs at this extremely important question? Why can't science give us a straight black-and-white answer and explain the line between protectable humanity and discardable tissue? Plenty of research money has been thrown at developmental questions these past years. And the tools of science are knocking on the doors of these ethical questions and begging for entrance. Shouldn't science say something about the difference between human tissue and human life?

The answer, regardless of how we feel about it, is no. Utterly. Absolutely. Inexorably. The reason for this negative answer is as straightforward as our anxiety about its silence is telling. We are asking science to make a judgment on something it cannot answer. Deciding the presence of human life in human tissue, under any circumstances, is not a scientific question. It is a value judgment.

Science cannot render an opinion about the worth of a fertilized human egg. It cannot ask what the speed of that worth is in a vacuum. It cannot cut that worth in half and see if it will grow back its other half. It cannot mix the worth of a salmon egg with the worth of a human egg and look for a physical interaction between the two. The same is true about deciding sentient humanness by looking for the appearance of certain familiar anatomical structures. So is asking science to confer a value judgment on what trait best describes human uniqueness. Science can measure only physical things.

Pride and Prejudice

When we consider all this complexity, we come full circle. Science hasn't produced the ideas of life and death, humans and tissues. We as a species have utilized these partitions to organize our environment for a while. But as the technology gets clearer, we are able to ask better questions and, just as easily, come up with more confusing answers. We find that older ways of organizing our environment are not as useful because they do not take into account new data. So we have to come up with different ways of arranging what we see. That's okay. It is part of the beauty, and the challenge, of new ideas. But we have to watch out for misunderstanding what new ideas are capable of saying to us. Or misinterpreting the power of observation in the midst of the observed.

Even though science cannot address this question, many people out there believe it can. And if for some reason science isn't performing up to snuff, one needs only to give it a little more time, and it will somehow save us. That prejudice has always amazed me. As a scientist, I have wondered why folks would bother to use research to buttress any opinion. After all, we are the same people who, over the years, have given you the atom bomb, toxic waste dumps and the ability to hear Geraldo Rivera nationwide every weekday.

The Power of a Good Reputation

Science has a certain amount of authority because its track record has fundamentally changed the way we live. We have sent people to the moon, found cures for diseases, and gotten "Nova" into homes as easily as talk show hosts. And science is run by people who attempt to explore things like objective reality. Objective reality has the ability to pull us out of our prejudices and attempt to put us on a higher mountain of truth. We'd rather conform our attempts to organize our environments in the company of people who subscribe to this kind of standardization than with those who say, "If I think it's true, it must be correct."

This yearning for an objective opinion is true of physical questions and, I think most intriguingly, also true of moral questions. We take as much comfort in knowing we have made a rocket for the right reasons as we have for making it at all. The operative word here is *right,* which means—or should mean—correctness or oughtness of a particular action. Many, many philosophers have attempted to crystallize this moral standard into a particular series of behaviors that can be shared by all humankind. Many have come to wildly different conclusions, but all of their ideas are treading water in the same intellectual ocean. That is, they seek some impartial standard, some objective but compelling reason, to behave in a certain fashion. It is the only thing that can elevate a prejudice to a Nobel Prize or pour an ethical idea into cement overshoes.

Discretion Is the Better Part of Value

But things can get blurry very quickly. When we talk about the difference between humans and horses, we seem to be addressing an issue at once scientific and ethical. Thus, it can be extremely difficult to see that such issues are still fundamentally different, even in questions where they appear to be superficially similar. The trick is to remember that they are different. That even though science and ethics both feed at the same objective trough, they do not digest their intake in a collective stomach.

These disciplines address such different aspects of any given question that their conclusions become nonsensical if they are placed in each other's realms. I don't mean to downplay this marvelous gift of research that God has given us; science is a wonderful philosopher's stone, transforming our leaden ignorance into golden nuggets of fact. The point is to understand that not all ignorance is leaden and that not all that is valuable is gold.

To the Jest of Our Knowledge

What does all this have to do with the little blue-eyed girl's question? I'd like to say that it has a lot to do with it, but in the end, I'm not so sure. Putting some of these thoughts on paper from such a sweet little inspiration has been a good experience. We are only in the beginning stages of understanding the rudest bits of genetic interaction in developmental biology, however. So the best shot I can give it is to say, "Here is a tool that many people think can answer that question, but it really doesn't." I'm sure that's an answer about as satisfying as drinking warm root beer.

In the next chapter, I'm going to address personal opinions about the subject and perhaps become more committed to a certain point of view. For now, I am committed to an idea that science can never perform a value judgment over physical phenomena. Perhaps later, when technology gets more mature, we will see more clearly the limitations of this twentieth-century god and understand that it has the same warm opinion about us that it does about flypaper. When we more fully comprehend the inadequacies of such false divinity, perhaps we will more fully understand the strengths of the real one. And then maybe my friend's little blue-eyed girl will be able to frame her question more succinctly. When it forms in her mind, she will only have to ask it in my hearing, and I will try to write her another book.

Chapter 11

Did Jesus Die for DNA?

I will try in this last chapter not to skirt the issues of human life and human death by describing why they can't be defined scientifically. Instead, I wish to commit as much as I can to an opinion on these issues with enthusiasm and vigor. To do that, I must examine sentience from an extremely personal point of view. In the beginning of this book, I said that most of my ideas concerning the value of life are derived not from the science I do but from the God I perceive. As one might expect, this chapter is religious in nature; it describes some of my attempts to understand the nature of a loving Jesus in the arms of an often baffling creation.

These musings run counter to the beliefs of many of my colleagues, a number of whom do not believe in a God, let alone a Christian God, let alone a loving Christian God. But since matters of faith have the ability to address value judgments where science does not, my opinion is most properly formed in this arena.

I feel very strongly about the value of a human being in the sight of this divinity. I would thus like to start this discussion by first addressing why I feel this is so critical, both internally as a scientist and externally as a cultural imperative. I also feel strongly about the general biblical ability to embrace scientific issues, specifically regarding this very touchy biological subject. I would like to describe my journey through several Scriptures that have, at various times, pierced me and confused me. I would like to conclude with a discussion about how these religious beliefs careen into what I perceive about genetics—and even try to air a few gripes about the collision.

When the Paints Go Marching In

The interface between science and religion is a hostile border to many of my friends and coworkers, who have at times expressed surprise—and worry—that a person who carries the badge of a research molecular biologist could also carry a faith in Jesus Christ. I am at once challenged by their questions and grateful for their concern. It is probably most obvious to them that if I address bioethical issues from an internal point of view, I must also address this most central part of my life. It is an extremely unscientific area, as I suppose all religious beliefs must be. But it is also the part made of the perceptions I most deeply feel and, in the end, of which I am most utterly convinced.

Let me start the whole thing by describing a conversation I had with a man who knew a lot about, of all things, stocks and bonds. He was a gregarious businessman, and one day at lunch he decided to tell me the story of value transfers. I had no idea what they were. But our conversation remained interesting because he used no financial terms that could get in the way of the communication, which was nice because it meant that I could understand it. He began his discourse by telling me a story.

"You know, John," he said, "the story of value by opinionated transfer is an old one."

"I haven't the slightest idea of what you are talking about, Bill," I replied.

"That's all right, John. I have an analogy for you. Let's say you were a portrait painter," he said, to my instant laughter. "Use your imagination, then," he said, also laughing, hopefully with me. "Let's say that you were selling your paintings on the sidewalk. Lots of people were milling around, but absolutely no one was buying. Maybe someone would offer you $5.00, maybe $25.00 if you were lucky. You were making very little money and were feeling pretty low about yourself and maybe even about your talent. Let's further say that it was at this low point you heard a commotion in the crowd. The people were buzzing because a famous painter was coming by, perhaps Rembrandt himself, coming over to view your artwork."

"Which would be impossible because Rembrandt is dead."

"Yeah, right, John. Nonetheless, let's say that he is coming over and you begin to tremble. The people gather around Rembrandt and fall silent as he carefully studies your paintings. After a few grunts to himself and a quick scribble on a scrap of paper, he says, 'This work is brilliant, mature, full of the lifeblood of a true artist, and worthy of a great price.' Rembrandt tells you to change the price of your paintings by tenfold and then buys one at the new higher price. What do you think is going to happen?"

I thought for a second. "The rules have changed, probably. Because of Rembrandt's opinion, the work has become more highly esteemed."

"Exactly," my friend replied. "And as a result, a higher price has been put on the work. Your paintings have, by the opinion of one man, become more valuable. That is what I mean when I talk about value transfers. Because of the importance of one man, the opinions that flow through him have the ability to perturb the value of things around him. Worth has been transferred."

He went on to talk about the stock market. The opinions of some people are valued so highly that the worth of stocks can rise or fall solely as a function of their opinions. It is intriguing to me that these examples of transfer can occur by esteem. It has been even more intriguing to find out they can occur in more places than the art world and financial institutions.

The Origin of Human Values

There are examples of this value transfer idea in certain moral questions. For instance, we can relate with surprising ease to people we barely know. We are horrified by what happened to the victims of Nazi medical experiments in World War II, even though many of them died before we were born. We sympathize with people who get beaten and left for dead, even if they are tucked inside the most furious corners of our urban problems. We project how it would feel to be in their shoes, shudder at the thought, and bingo, transfer a bit of our own worth right on to them. Sometimes I think we do it to convince ourselves of our own worth. If other people have value, and we are a lot like them, somewhere along the line we have value, too. Additionally, I think that we do not wish to become cold automatons, as if being a cold automaton were intrinsically bad. Instead, we wish to be loving, value-laden human beings. Whereas we might speculate on individual motivations, exactly where those ideas originate collectively can be a very troubling question when examined closely.

For example, we in America might say that it is morally wrong to kill people on the sidewalks of Tokyo. Why is it impermissible to kill people on the sidewalks of Tokyo? We say it's wrong because they have individual worth. And where did the people on the sidewalks of Tokyo get that individual worth? Who gave them quality badges of such net intrinsic value that we would spend millions of dollars on one of them if he got caught in a mine shaft? There is no reason necessarily for them to have worth. Looking naturally, aside from the ability to identify with them as fellow human beings, there is no a priori reason to care about them at all. Or anybody else you do not know. As a geneticist, I can tell you that there are no such things as worth genes. Financiers will tell you there are no human worth exchange rates in the money markets of Switzerland. We wouldn't spend millions of dollars on a sidewalk Japanese if he or she were a homeless beggar. But if the person's life were threatened—especially in a publicizable fashion—there isn't enough money in the world to lay at that person's suddenly invaluable feet.

So we still continue to value human lives. Some of it is undoubtedly due to training; some of it is due to our ability to identify with similar experiences. At this point, some of these opinions get lost in a mass of intriguing cultural values.

And in my opinion, a single magnificent divine one.

Moving Heaven and Worth

I've stated before that in social situations, I have been asked if knowing about genes and chromosomes changes the way I look at—or value—human life. This directly relates to an opinion concerning the sentience in human cells. I have stated in those social situations that the miracle I believe God has wrought deep in our genetic underpinnings has made me cherish human life all the more. And as a function of the study of that life, I have become more enamored of the One who created it.

This view doesn't detract from my perception that predictable, normal biochemical reactions power the engines of such biological miracles. But in the end, I don't cherish the molecules and chemicals. My mind reels when I consider the incredible mind and blindingly creative brilliance that breathed those molecules into existence and gave them motion and color and reproducibility in the first place. While I am in awe of the design of complex living things, I am absolutely speechless at the greater complexity of the Creator of living things.

I thus believe extraordinary differences exist between dead human beings and live ones, differences that transcend the complexities of their common, though quite opposite biologies. To discern wisdom in the middle of those differences is to find the heartbeat of God in the center of otherwise decadent flesh. To me, His attitude toward us is at the epicenter of human worth. I would like to use the idea of value transfer to talk about how it arises.

Dust in the Wind

Where do humans derive their worth? I believe that humans derive their worth from a transferred opinion about their potential destinies. I think this opinion is transferred from a loving God to generally not-so-loving humans. Why do I say it is transferred? Mostly because I believe that aside from God, our biology is nothing but a finely tuned bag of genetic chemicals. I am convinced this is the reason we can construct biological life so easily in the laboratory. Biological life is simply molecules, flexible enough to make a palm tree or a tiger for sure but, in the end, purely atoms. And that's all Scripture says we are, too, a rather indiscriminate pile of dirt. Although that dirt may be a fascinating source of origins, as far as our bodies go, from dust we came and to dust we shall go.

Almost. We are left with dust and destruction only if we have no God to energize our existence and give us reasons to believe in more than our philosophical belches and our ability to wage war on each other. I additionally believe, and consider it fortunate, that dust is not all that lies within the human potential. The difference lies in the fact that something was transferred to us as a species that was not transferred to other created things, such as rocks and butterflies.

I believe what was transferred was divine in origin and thus inescapable in application. God breathed something into us; God breathed His image into us; God did something to us that transmitted value to us. This simple bending of His love toward us gave us all of the worth we possess. Which is infinite, considering who was doing the bending. What we became as a result of that attitude was something of a gift; even in rock bottom carnality, we were transformed into creatures the Bible describes as only a little lower than angels. We became eligible for an eternal glory, a bag of molecules suddenly given the indescribable worth of a collective crucifix. As a race, we were, and indeed are, the recipients of a tremendous degree of value.

This worthiness is not just collective, however. It is as individual as the numbered hairs on our heads. And since that number changes every day, so does the assessment, which simply means that Someone is paying attention to us all the time. The transfer of worth is extremely personal. Particular. As individually tailored to our biology as is our exclusive genetics, both oddly seeking to ensure absolute identity in the face of hostile pretenders. To me, this individuality is equally terrifying and emancipating; frightening because it is so intimate and liberating because it is so full of love. It is amazing that someone who is so incredibly talented thinks so much of us. Or wants us so close to Him.

I will always be impressed that this worth tells us a lot more about the Giver than it does about the receiver. But whatever the interaction, the point is that worth is transferred. Like the paintings of some artist being inspected and affirmed by the hands of a master. However, unlike a human artistic genius, we do not have to rely on a subjective opinion. We possess the worth given to us by Someone who, if I read the Bible correctly, has the ability to do anything He chooses perfectly. And He has chosen to value us.

Pushing and Loving

Perhaps I say all this to affirm that scientists can have religious beliefs and still be scientists. Yet my placing the value of human beings squarely on theological shoulders hardly settles the issue about the beginnings of human life. You can ask several important questions: If a human being is established when that worth is transferred, when does God transfer it? To what does He transfer that life? Does God transfer worth when He sees a human as an individual biological entity? Does He transfer that worth at conception? What about at other times, like when the brain is at a certain stage of development?

In the face of these questions, I must turn to the Bible as a source of inspiration and guidance. As a Christian, I hold the divine authority of the Bible to be axiomatic. For issues clearly spelled out in Scripture, that authority is simple to comprehend and can be straightforwardly—if not always easily—installed in one's life. There are problems in understanding its authority over issues that are not clearly spelled out, however. And even greater dangers accrue when the Bible is forced to make decisions over things it was never designed to comment on.

Why the Bible Is Not an Organic Chemistry Textbook

One example of this misuse comes when the Bible is treated like a scientific textbook. When that happens, we leave it open to probe questions it was never designed to address. For instance, one cannot look to the Bible for a list of Newtonian physics formulas. Neither Genesis nor Revelation contains lookup tables for the genetic code. One cannot find a divine unified field theorem in the Old or New Testament. This inability does not weaken the Bible's ability to direct the lives of humankind. But just as I cannot find the road to peace with God in an organic chemistry book, I will not find the answers to specific technical questions in Holy Scripture. That is not the intent.

The power of this mistake is seen when it is applied to certain biological questions. I have stated that the differences between tissues and life are not scientifically answerable questions. Yet I have been to a number of Christian seminars that claim these questions are addressable scientifically and THEN there is an attempt to use the Bible to buttress this claim. Two mistakes are made: giving science much more power than it is capable of delivering, and misrepresenting the authority of Scripture in our lives.

I believe that the Bible is absolutely silent on the issue of the beginnings of human life. I also know I am very much out of my league when I attempt to address topical biblical scholarship. Nonetheless, as a scientist, I am keenly aware of how others have used certain passages of the Bible in an attempt to validate personally held points of view about science. The purpose of this section is to demonstrate this biblical silence and to see what happens when we treat the Bible as a scientific text. We will specifically ask the Bible to tell us exactly when God sees human tissue as human life.

A Scripture that always comes to mind concerning the status of human life is from Psalm 139:13–14:

> For you formed my inward parts;
> You covered me in my mother's womb.
> I will praise You, for I am fearfully and wonderfully made.

I believe this Scripture says that God had a direct hand in the formation of an individual personality. It appears to demonstrate that this personality was recognized as such in the uterine environment. An individual is thus resident within the womb. We are tempted to conclude that God has injected His valueship at least by the postimplantation stage of the blastocyst. From then on, while it is in the uterus, it is a personality that God recognizes as unique and worthwhile; as such, it cannot be tampered with. We know nothing of the preimplantation stage in this Scripture, nor do we really know anything more than that God recognized the author (David) as unique. Without further information, whether He actually recognizes anyone else as unique is an open question.

Jeremiah and Jesus

If that is all the Bible said about human life, we might safely surmise that life existed at least when the little embryo carved out a nest in the uterus. We would then be left to speculate on our own about what happened in the preuterine environment. Or about God's opinion concerning people other than David. So we look for other Scriptures, seeing if we can find other tissues and other characters that God might say something about. In so doing, we find Jeremiah 1:4–5: "Then the word of the LORD came to me, saying: 'Before I formed you in the womb I knew you; before you were born I sanctified you.' "

It seems that God knew about an individual before the embryo took up residence in the womb. Or at least before the embryo had formed in the womb. That is, value was added to a group of cells before they implanted and started developing inside the uterus. So now we are back in the Fallopian tube. This might mean that the Bible is being very thorough about its definitions of human life and is including the preuterine development in its examination. From this, we might conclude that God discerns a unique person prior to an embryo's implantation into the uterine wall; God has injected His value somewhere in the first days of life, answering the vacuum left by the previous Scripture and creating another all on its own. Under these conditions, we should not be meddling with any embryo after it has been conceived and before it has been implanted. Doing so would be a violation of God's vision for the person. At least as it concerns Jeremiah.

If we take these two Scriptures together, individuals appear to be identified in both the uterine environment and the preuterine environment. How about at conception, when the egg and the sperm fuse? Events in the Bible infer, although they do not describe, an instant when God appears to recognize individuality at cell fusion. For example, Jesus was completely human and completely divine the moment Mary became pregnant. Since Jesus was a He, the Y chromosome that made Him must not have had a human origin. This invasion of genetic material did not occur when Jesus had a brain wave. It did not occur when He possessed a heartbeat. In fact, it did not occur when the fetus crossed a developmental milestone. Instead, the promise to Mary occurred at the moment of conception. God apparently recognized Jesus as an individual human personality the instant a divine germ cell united with a human egg.

People have inferred that because of this recognition of Jesus, the recognition of all humanity must fall under the same rules. Although that is an interesting leap of logic, it provides some stability; if one combines Psalms and the book of Jeremiah in context with this idea of a virgin birth, a scriptural framework is established to answer the cell/sentience question. What is derived is that God discerns protectable humanity in both the uterine environment and the preuterine environment because He sees protectable humanity at the moment of fertilization.

Levi's Genes

If that was the final word the Bible pronounced over the presence of sentient humanity, we might have a scriptural basis to outlaw experiments or manipulations with any fertilized egg. If God recognizes preimplantation embryos as human beings, we should not mess with them in any shape or form. It does not matter that they are potential newborns and not actual newborns; if God recognizes them at a certain stage, we should, too. This would be quite a comforting and explicit answer to the question about life and tissue. The problem is that other Scriptures comment on the presence of human individuality before the preuterine environment. Consider these troubling verses: "Even Levi who receives tithes, paid tithes through Abraham, so to speak, for he was still in the loins of his father when Melchizedek met him" (Heb. 7:9–10).

These verses appear to recognize an individual human being before conception. The genes that will eventually form Levi are still in Abraham's testicles, which means that Levi was recognized as a unique individual before he was a fertilized egg. Whoops! So much for life at conception. Or even sperm formation because, in the testicular environment of Abraham, Levi was a partial collection of genes. Abraham was not Levi's daddy. Jacob was Levi's daddy. How were Jacob and Abraham related? Jacob was Abraham's grandson.

Since sperm cells have a half-life of sixty-seven days or so, the only person who could have contained the sperm encoding half of Levi's genes was the grandson Jacob. And Jacob could have carried the "Levi sperm" cell for only two months prior to the act of intercourse that would establish Levi's womb experience. Abraham's sperm cells could contain only bits and pieces of genetic information that generations later would become Levi.

This Scripture, examined rigorously, appears to say that God recognizes future genetic information as distinct, protectable humanity before formation as eggs and sperm. Which means we are in big trouble. If we are to use the same rationale concerning potential humanity as in previous Scriptures, we better save the product of every ejaculation on the planet. And then we better deep-freeze the ejaculates for at least three generations, the minimum tenure of God's recognition, according to the book of Hebrews. Not to do so would be a violation of the divine transfer of worth. Why? Because as this Scripture points out, God recognizes individuals as distinct human beings before they are germ line cells.

Dirt Feelings

As we can see, using the Bible as a scientific guide to establish the definition of unique personhood can lead to rather odd conclusions. But the Bible isn't finished with us yet. We must contend with another, even more puzzling, Scripture if we are going to use Scriptures to tell us when sentience begins. Read Psalm 139:14–15

> I will praise You, for I am fearfully and wonderfully made;
> Marvelous are Your works,
> And that my soul knows very well.
> My frame was not hidden from You,
> When I was made in secret,
> And skillfully wrought in the lowest parts of the earth.
> Your eyes saw my substance, being yet unformed.
> And in Your book they all were written,
> The days fashioned for me,
> When as yet there were none of them.

The lowest parts of the earth! All of a sudden God is looking at the lowest parts of the earth and seeing an individual. You can ask some interesting questions if this phrase is taken to mean terra firma. Are there people being wrought in the depths of the earth? Is God saying that human pregnancy isn't in the womb but in a deep dark cavern? The text states, after all, that the writer of the psalm was made in secret. Is the

Bible discussing a hidden form of human biological development? My opinion is that the Scripture is probably not saying that at all. I believe there is a more subtle idea at work here. But if that idea is taken as a scientific commentary, it leads to a ridiculous conclusion.

When one examines the depths of the earth, one finds carbon and nitrogen and hydrogen and sulfur and phosphorous, the very chemical constituents of biological life, which includes, of course, human life. These atoms and molecules will be used to form human beings in a natural food chain. Some of those molecules in the ground will be absorbed and eventually become parts of plants and animals. A human will eat one of those plants or animals and, in the normal part of digestion, insert some of those molecules into his or her tissues. A few will be shuttled off to become parts of the DNA in human sperm or egg cells. And eventually those cells will become part of a pregnancy. So the molecules in the dirt can be traced directly from their terrestrial origins all the way into a birthing room.

This Scripture appears to say that God sees an individual personality while that personality is a random collection of molecules in the ground. In this case, God sees an actual individual while that individual is a mud slide. Or a pile of dirt. Does that mean that mud slides and dirt piles become invaluable and priceless entities because God can recognize a human in them? Following the rationale we've discussed, we now will have to save all the dust we encounter. Or create giant mounds of dirt in our backyards and be careful who walks on them. Why? Because if we take this set of Scriptures in its most scientifically literal meaning, every dirt molecule is a potential bit of a God-recognized humanity. And we will have to value this substance as we value any substance that God recognizes as human, whether it be in the womb, Fallopian tube, or testicle.

The Deafening Silence

At this point, we must begin to think a little bit. If we follow the rationale that everything God recognizes must be treated with the same dignity of normal adult humans, we will start with protecting infants and end up protecting dirt piles. So either those Scriptures are dead wrong, or they were never supposed to be used to interpret the beginning of life. Historically, those Scriptures did not have anything to do with the definition of human essence. They were mostly used to describe the predestination of the believer. And if all the Scriptures we were using to define sentient life mostly address predestination issues, the Bible is silent about the specific instant of the beginning of life.

I have a love/hate relationship with the lack of scriptural direction in this issue. I have always been tantalized by God's ability to share His creation with other individuals, such as ourselves. I have been even more amazed that God could create a creature like a human and give the being the full ability to study this creation in an independent and thoughtful manner. I often run into these feelings in the laboratory. I am frustrated when I perform an experiment that doesn't work, and I am exhilarated when I discover something that was previously unknown. And that is the source of the love/hate. Because I am a Christian, His silence fills me with a sense of wonder, sometimes genuinely ominous, sometimes genuinely exciting.

Because I am a genetic engineer, however, this silence drives me crazy.

Worth Defects

So what do I think about the differences between human beings and human tissues? Up until this point, I have mostly discussed the outline of a biological vacuum, in essence describing the boundaries of nothing. We have bumped into interesting border guards in this safe description; there is science, which is at a loss to explain these differences, even as the Bible, which can explain them, chooses not to. We are left with a divine enthusiasm for human life—provided you believe in the same divinity—and an incredibly complex mechanism upon which He confers His love. And that is just about it. How can we use these ideas to derive the sentience in human cells? Are we left with anything that could tell us what the differences might be between human tissues and human beings? Are we really expected to figure out where London is by first exploring the corners of the earth where it is not?

I have so many opposing feelings regarding this issue that most of the time my common sense gets lost in an ocean of contradiction. I am at once untethered from scientific inquiry even as I am bewildered by biblical interpretation. I have the sense of living with an eerie professional—and extremely personal—ambiguity. In other words, I don't know how to answer this question. I don't even have a good idea where to start. So my first inclination is to pack up my word processor, burn the lab bench with all its attendant commitments, and go sell flowers on the street corner.

But only for a few moments. There is always the hope of God's creativity even in the midst of human dullness. I have recently begun to understand that confronting the very dilemma does something to weaken its potency. This confrontation shows there may be a framework for discussion in the midst of the chaos—kind of like discerning the dim outline of the trenches from the smoky middle of no-man's-land.

Let me start out by discussing a few things that I cannot get around, no matter how hard I try. Perhaps value can be derived by defining the boundaries of this conflict, even if I cannot tell who has won the war. Some of these struggles concern what it means to be a religious person in the world of research science. And just as personally, what it means to value a human life in the grips of a powerful genetic discipline.

School of Hard Clocks

I'll first air a few gripes. One argument that really drives me crazy is the idea of potentials. Basically, it is stated that the potential abilities of a given set of human chemical reactions comprise the fulcrum upon which we judge protectable humanity. I always wonder how potential activity says anything about internal composition. It's like saying the potential earning capacity of a spin at a roulette table says something about the kind of wood from which the table is constructed. From a tissue versus life point of view, this is nonsense.

A teenage girl may be killed and never reach her biological potential. But no matter when she dies, no one will doubt that when she was alive, she was a human being. The same is true of a newborn girl. She may die of a number of childhood diseases. But when she could follow certain movements with her eyes and laugh and gurgle, she was a human being. Neither person in death realizes all the potential she might have had in life. But what does that matter? This failure in no way changes how we feel about the residency of the girls' humanness. There is no qualitative difference in their overall humanity simply because of their length of life.

I believe that the same is true of the difference between the newborn girl and a fertilized egg. As we have discussed, a newborn has a clock within. Given proper conditions, the newborn under the watchful eye of this clock will turn into that teenager. So will the fertilized egg in the reproductive tract of its mother. It, too, has a clock, wound back only by nine months compared to the newborn.

What is the difference between the two? To me, it is merely a difference in the time spent on the planet. This tenure does not describe a basic difference in the original derivation of the organism. From a scientist's viewpoint, creating a biological distinction based on such time is capricious. It says nothing about the internal composition of the biological entity in question. I continue to feel that the argument of potentials is a squabble over irrelevancies. The road linking the potential with internal composition is not only lightly traveled, it is imaginary.

Fetal Attraction

One might be tempted to disagree, perhaps strongly, with this statement. One might say that time is extremely important, rather than arbitrary, in considering the composition of human development. After all, the embryo is constantly changing its internal cellular constituency. Each change is a function of time. Thus the moment at which you examine an embryo is significant in discovering what you will find. As such, time is not imaginary but is instead a compelling and critical part of any definition of human life.

This argument always puts me in a bad mood—and leads me to another one of my gripes. These ideas inescapably imply that a human being could be something other than a human being at certain stages of development. I think this is nonsense. A fertilized human egg can never be anything other than a human being. Consider for the moment that the only nonhuman-being participants are human sperm and human eggs. These are individual human cells, with the main distinctions that they have no clocks ticking inside them that in a certain period of time force them to become human beings. If you put only an egg or only a sperm in a uterus, the only thing you will get nine months later will be an empty uterus and a dead germ cell.

A fertilized egg in the female reproductive tract is a very different story. This tiny combination of cells can produce a little baby. And most important, it will not—indeed it cannot—produce anything else. A fertilized human egg cannot be a monkey being. It cannot be a sea horse being. It cannot become an anything-else-being because it is a human being. From my point of view, its uniqueness is defined not by how it functions but by what it is. And the time spent in the uterus tells me only how small that human being is, not whether a human being exists.

Understanding the importance of the clock is a critical issue with me. But not for just personal reasons. Cherishing the value of any human life has consequences that extend beyond an attempt to satisfy certain internal standards. This absolute value forms the glue that binds most of us to Western civilization. I believe this value has affected our political thinking, our civil codes, and the way we view our neighbors. I think it affects how we fight war, a time when we otherwise throw every moral instinct out the window. But it's a safeguard for an extremely important reason, and its argument constitutes another gripe of mine.

A Clean Bill of Wealth

Something that is not human can always be treated as property. This is seen all the time in the burgeoning industry of biotechnology. We can patent certain discoveries from human tissues as legally defendable pieces of merchandise. We can take molecules, derived from humans at all stages of development, and market them like laundry soap. So one way to look at this question is to say that we are attempting to define the difference between human beings and property.

And this redefinition leads me to an observation. I have never met anyone who is a certain percentage human being and a certain percentage property. It is impossible for me to think how this could be achieved. People who are mostly held together by orthopedic pins and needles are required to pay taxes. A human being may possess property, whether it is furniture or part of his or her own cellular resources. Although you may take away huge percentages of furniture or tissue, the person is a full-fledged human being if the person is still alive. Not a subhuman being, so named because he is missing some of his biological resources. We must be willing to treat the person with all the rights and dignities of any, less-damaged human. Such an idea constitutes the moral basis for many of our laws concerning persons with handicapping conditions.

In addition, a person is not judged as a continuum of value, with property constituting one end of his personhood and humanity constituting the other—and lots of intermediate percentages between. The person exists, or he doesn't; otherwise our obligation to treat him in a humane fashion would be a function of the percentage of his humanity. Such a distinction, in my opinion, is rubbish. We do not, nor can we ever, have fractions of people. We can have only fractions of value that we place on them.

I say these things not as a specialized scientist but as a generalized member of our species. As I have stated before, these conclusions cannot be arrived at experimentally. They are arrived at through a process of continual value assessment and subjective opinion. In my case, they are also susceptible to the influences of my religious beliefs. As strongly as I might feel about them, however, I must be continually aware that these opinions about human life have one foot in an open mind. They must be rendered flexible enough to accommodate new data, whatever that data might be. And from wherever direction it may come. Even, or perhaps especially, if it comes from genetic engineering.

The Problem with Gray Matter

I have often wished that every issue I encountered in biology, whether it be medicine or genetics, would consist of a single choice. Like a computer, where something is either on or off. Soon my outlook on life would be a monochromatic universe of black and white, which would be the point. Under such conditions the decision to paint an issue a certain color would be dramatically simplified. And if God had chosen to paint the canvas of human life with such duality, the argument about its essence would be lost in the light—or the soot—of a solitary obviousness.

The problem, of course, is that human life is not a simple thing. The palette He has chosen to color us with is as infinite as His creativity and as profound. The decisions we make regarding its manipulation must take into account every hue He has infused into it. The fear of leaving something out—and missing His entire intent—grips me strongly when I view His masterpiece under the microscope.

As both a scientist and a believer, I often struggle with understanding the difference between absolute skepticism and absolute certainty. I have always confronted it awkwardly; the battle often permits me to see both sides of an issue, without allowing me to make a decision. The differences between human tissue and human life constitute one such issue. For example, the ideas enunciated in the last section place me for the moment squarely in the arms of the life-at-conception camp. But this decision does very little for me when we brush the borders of certain genetic exceptions. And here comes the nature of the conflict. Conception and, indeed, the value of human life are not always monolithically definable processes. We can give examples of humans that look familiar to us but may be merely empty biological shells. And it turns out there are ways to create complex animals, even mammalian embryos for a period of time, without bothering with conception at all. As a scientist, I am forced to deal with these questions, even if as a believer, I am forced to deal with their answers.

Exceptions to the Duel

Let me commence the description of this ambiguity by airing another gripe. Something that really gets my goat is that we as a society choose to have differing standards about the sanctity of life, especially when we consider the entire family of humankind. In World War II, we might have defined an enemy soldier as a human being from the start, but we did not necessarily confer on that human being the right to live simply because he existed. Many persons who violently oppose the destruction of single-celled human beings feel perfectly justified in bombing to smithereens multitrillion-celled young men and women. The only distinction is that these bombed-out men and women possess different political or religious persuasions or have different leaders egging them on. This lack of absolutism is seen, and perhaps most keenly felt, at the personal level. Most of us would likely use a gun on someone about to murder our own son or daughter—even if we swore up and down that we were strict, unshakable pacifists.

By the situations mentioned above, we have long since made our standards not absolute but conditional. In so doing we chose the slippery slope before we even saw the hill. The point is that the sanctity of life, while in principle quite noble, is not an inviolate code in every human value. It is an idea that appears to have many, many exceptions. At what point do we apply those rights to someone who has yet to draw breath and may be incapable of doing so?

The issue of the anencephalic child looms large in my heart. As you recall, this fetus is born without a cranial vault, missing most of the brain. Here is a creature I would call a human being in its most vulnerable and least defensible configuration. Even though it is a human being in my estimation, I am not yet prepared to give it all the rights of other, less-deformed human beings. Why? Because defining something and then sticking rights onto it like postage stamps are two separate things.

The presence of the anencephalic baby brings up many issues when one considers the sanctity of human life. Perhaps the most glaring issue is the validity of using thought-life in the primary definition of human sentience. That seems funny at first, because human thoughts make such readily apparent markers for the presence of personality. And it is convenient except for two things: No one has defined what a uniquely human thought is, and everyone who has been involved in deciding when human thought is absent has called the decision arbitrary. We don't even know the biochemical difference between a reflex and a deduction.

Whether we are willing to give anencephalics all the rights of nondeformed human beings is to me an open question. The problem would not be nearly so sticky if there weren't benefits that could be derived from their use as organ donors. We must ask whose moccasins we are willing to walk in. If I had a baby girl who would die unless an organ donor was found—and an anencephalic donor was available—would I use it? Is this question much different from saving the life of an offspring in the face of criminal activity? I have not answered these questions in my mind. I am, in many ways, still contemplating the effects of applying different standards of worth to organisms I define as human beings.

Coming Apart at the Dreams

The future has always weighed heavily on the present with certain kinds of research. Kind of like tomorrow coming down from obscurity for a minute and going "boo!" Usually such pressure is the office of the past. But to understand part of this dilemma, we will have to look a little farther than we can at present see. And to do that, we will have to peer into the crystal ball of the future of genetic engineering and determine if monsters are looking back at us.

As a practicing researcher, I tend to resist the temptations—internally as well as externally—that urge me to predict the future of genetic engineering. In a field where emotional reactions become conspicuous and realistic insights hidden, truth tends to get sacrificed on the altar of innuendo very quickly. And it gets worse when one tries to prognosticate in a field that changes with such velocity. Quiet introspection can be interpreted as imminent disaster if enunciated publicly, which has a habit of mutating into solid fact without the benefit of experimentation and hard data.

I say all that as a caveat to the dangers of speculation because, in this section, that is exactly what I am going to do. There are experiments, already performed, that I believe will lead inevitably to certain genetic capabilities. These capabilities will bring into sharp focus moral questions and ideas that may lay an axe to some of our traditional notions regarding human life today. Let's begin our discussion with some ideas we referred to in chapter 6, that is, the molecular basis for human conception.

If in the anencephalic question we can still call something human, we are at least comforted with the idea that the child was formed by conception in a natural, normal way. But what happens to the placement of our values when we behold organisms that come about in ways other than by conception? One idea that directly challenges our notions of human life comes from what we find when we look under the hood of this fertilization. As we have stated, molecular biology, working hand in hand with its older sibling embryology, is beginning to demonstrate a troubling fact: the collision of two cells is not some all-consuming, can't-do-without-it barrier by which complex organisms must be made. Instead, it offers us alternatives for creating organisms, from oddly placed embryonic fusions to parthenogenetically derived creatures. And even mixtures of the two.

Some of these results come about because of the attempt to understand the normal process of fertilization. In chapter 4, we likened the natural collision of egg and sperm to an exquisitely delicate hand touching a marvelous, preprogrammed switch. When the egg becomes fertilized, this switch is activated, setting into motion an intricate ballet of chemical reactions. This series of reactions is the developmental program, codified deep within the genetic information of the participating cells. When the genes of these cells are turned on in a proper sequence—undoubtedly interacting with factors already in the cytoplasm—an organism is created. What is important is not the collision of cells but the activation of this program. And if this program can be activated without the fusion of cells, this collision is not a requirement, but an option, in the creation of complex life.

I say it in just this fashion because there is a tremendous desire to understand how this program is initiated. Researchers want to know which genes are activated first, which factors in the cytoplasm are needed for that activation to occur, the contribution of the initial as well as the subsequent cells, and so on. One by-product of these experiments will be the activation of the developmental program without prior fertilization.

Or at least a partial reconstruction of the initial events. Otherwise we will not be able to say conclusively which molecules are involved in which step and which ones are not. Although the time frame before these results occur may be open to question, I believe the fact that they will occur is not.

Appearances Can Be Conceiving

We are already gaining clues about the triggering events from the manipulation of cells after conception has occurred. As we discussed, we know that individual blastomeres—those cells that arise long after fertilization has taken place—are just as good at creating a complete mammal as a fertilized egg. So are blastocysts, a group of cells formed at a later stage of development. Conception may be needed to get a certain embryo started, but you can take pieces of the embryo and make whole new organisms.

Why can this be done? Although the biochemistry is undoubtedly complex, the reason is fairly simple. Resident within these blastomeres are all the ingredients of the developmental program. And while good old-fashioned conception may be needed to create these cells initially, it will not be needed to sustain the generation of new organisms. In theory you could implant a blastomere into the reproductive apparatus of a mammal, wait for it to form other blastomeres, harvest a few of them, reimplant them into yet another animal, wait for them to form more blastomeres, harvest a few of them, and reimplant into another mammal ad infinitum. The animals possessing an implanted blastomere would become pregnant with genetically identical fetuses. You would end up with untold numbers of twins—all from different mothers. And more to the point, all formed without the benefit of conception. It's kind of like picking from a tree that is continually bearing fruit. Most of the work surrounding such blastomere transfer has been done in agriculturally important mammals, such as cows. But it is probably feasible in human embryos as well.

So what does this tell us about conception? The primary collision event is needed to get the ball rolling, but such an event is not needed for every complex organism created subsequently. Cells that contain a triggerable developmental program are required. One could still say that because we need an initial fertilization event to jump start things, life has its beginnings at conception. Embryologists in this case would be extending the productability of a single conception. And that's a good point. What's needed is the demonstration that complex life forms could be created without any initial collision of sperm and eggs. As we have discussed, we already know that this is possible.

A Fatherless Child

The process of parthenogenesis demonstrates that some organisms have learned to activate their developmental program without any initial collision of two cells. For this to occur, a cell that contains a triggerable developmental program is necessary. In naturally occurring cases of parthenogenesis, this cell is an egg. Certain creatures such as lizards and other complex organisms do not need—and in some cases do not have—a member of the opposite gender. This is not an experimentally manipulated piece of data. It is an observation of the life-styles of organisms that already exist.

But there is more to this story than the zoological observation. Parthenogenetic activation can be performed in organisms (mammals) that normally demand a collision between sperm and eggs to create an offspring. For mice, one merely requires the presence of an egg. The developmental program in this egg can be stimulated by dipping it into a dilute solution of alcohol. The implantation of this egg into a uterus will result in the formation of an embryo.

The ability to re-create at least initially complex organisms without fertilization is already an experimental reality. So far, an embryo has never been taken completely to term from a single parthenogenetic stimulation. The organisms that are formed usually die after a certain period of time in the womb. Some people speculate that a whole mammal can never be created from a parthenogenetic fertilization. Recently, however, researchers have been able to take half a mouse embryo derived from a parthenogenetic conception, fuse it with half an embryo derived from a normal fertilization, and get a complete mouse. Whether a complete organism will never be created from parthenogenetic stimulation, or whether it is only a matter of time before one can be created, is an open question. The point is that the activation of the developmental sequence has been established, at least in part, without conception.

No Clone Unturned

I have read, mostly in pulp magazines and tabloid headlines, about the ability to clone human beings from bits of their skin cells. This, of course, is foolishness. Disturbingly, however, the basic premise behind that fanciful notion is not foolish. As has been stated, all the genetic information necessary to describe a human being resides in nearly every cell of that human being. This fact is true of most organisms. The upsetting thing about such complete redundancy is that this genetic information includes parts of the developmental program. At least in theory, placing fully differentiated cells—say, skin cells—in the proper cellular environment might trigger aspects of that program in them.

This reasoning was the basis for the series of experiments described in chapter 6, where a toe cell from a frog was inserted into an unfertilized, enucleated frog egg to create an entirely functional tadpole. Does that mean we are capable of creating complex organisms from tiny bits of nongerm material? Certainly not. And I do not really know about the future. To be honest, that frog experiment—if true—says as much about the chemical information inside a burned out egg as it does about the genetic information within the toe cell. But there is a theoretical basis for the manufacture of embryonic cells from preexisting fully differentiated ones. And what it might mean in the future, perhaps the far future, is an open and quite unpredictable question.

Whether we are discussing the totipotency of blastomeres or the credibility of tabloid headlines, many of these ideas frighten me. We have just begun to tear apart the developmental program, isolating the important genes and discovering the mechanism of their ignition switches. We are learning how they interact with each other and with their immediate environment. Already we have made some breakthroughs, from the discovery of homeotic genes to the isolation of MyoD1 and its gene family.

In terms of time, I have to integrate these enormous strides into some kind of perspective. Real genetic engineering started in the mid-1970s. What has been accomplished since that time is nothing short of a revolution; we already are pounding on the door of the sequence of the entire human genome. And the field continues to move very, very quickly.

I believe that we will eventually find the ignition switches for the developmental program now mostly resident inside fertilized eggs. I believe we will eventually know the phone lines that hook up the cytoplasm to the nucleus of a fertilized egg and how that communication results in programmed cell division. I believe that we will eventually understand the sequence of events necessary to trigger the formation of embryos without the necessity of the natural players.

And because I sense that all this will happen, I have begun to question our traditional notions of human beings and human tissues.

These ideas do not necessarily obviate the importance of human beings in their tiniest, most vulnerable form. But these ideas do show that we may be able to generate them, or at least parts of them, in nontraditional ways. Which indicates we will have to rethink what the traditional ways really meant in the first place. If we learn to make parthenogenesis a human reality, unfertilized egg cells will be elevated to the same status as fertilized egg cells in terms of their potential for creating protectable humanity. That means we will blur the distinction between human beings and human tissues. If someday we are able to create living beings from toe cells, every cell in the body will become a potential human being. In which case all of our tissues will be elevated to the status of fertilized eggs. This to me is frightening because it means we will no longer have blurred the distinction between human tissues and human beings.

We will have obliterated it.

Absence of Callous

I have attempted to describe in this chapter a very personal wrestling match regarding the differences between human life and human tissue. As you can see, I have not come to many conclusions about their absolute boundaries. I am increasingly left with hints, especially as more light is shed on the molecular basis of human development. It is extremely uncomfortable to have such a vital issue with edges that appear so undefined. Or perhaps it is uncomfortable only because science cannot give me a tangible answer and I so wish that it could.

Science can never change some things. No amount of research will change for me the fact that God values us tremendously. Nor will it change the fact that we don't deserve most of it. His willingness to transfer some of His image to us raises our worth to the level of that value, and it happens simply because He loves us. I cannot help treasuring such an attitude from the heart of Someone who is so powerful. In general, understanding who generates such an odd balance of payments compels me to value tremendously anything on which that worth is bestowed. Which means, in the end, I must value human beings the moment there is a chance they might exist.

This willingness to transfer value distills for me the subjective part of this issue to these questions: What does God love? And when does He love it? It is to me a divine fulcrum upon which those very earthly questions are balanced. That's why it is no surprise to me that science cannot address these problems. Science doesn't believe in miracles. Not that miracles don't exist. It's just that science cannot measure them. Looking for divine transfer is not something you can write a grant over. I'm not sure it's something that can be described by a law.

I'm happy that this distillation provides an internal framework for my faith. At present, however, it is all I am capable of understanding. I know that God has not raised the issue much in the Bible. That's okay, really. The Bible wasn't written at a time when manipulating genetic material was much of an issue. I realize that it doesn't have to be explicit to be relevant. And it fairly shouts until it's hoarse about God's relationship with His wayward race. Because it is silent about the specifics, however, it leaves unattended a rather important vacuum. And the presence of this vacuum forces me to be open-ended about some of the borders of life and tissues. I suppose it doesn't help that research science does the same thing.

The Stature of Liberties

There is a problem in clinging to absolutes that not everyone sees anyway. Even if the Bible told us explicitly the answer to this question, not everybody would listen to it. At the same time we share a common biology, we as a race of people do not share a common perception of values. And that makes things amazingly awkward because we all possess genes and chromosomes that are organized in such a fashion that we also possess sentience. Most of us can reproduce this sentience in the form of children. But we do not hold the same opinions about this sentience, not in its essence, its beginnings, or its endings. And the attempts to conform all of humanity around a central idea make those who don't believe in such centrality angry. That's the problem when this issue is perceived in a subjective fashion. And because science can't validate subjective fashions and because many people don't believe in absolute standards outside science, we have a giant conflict. It ends up becoming a battle of mutual intolerances, not a discussion of mutual interests.

To be sure, in the midst of these many voices, I have my own prejudices. Fueled perhaps by my thoughts concerning human value, I choose to place tremendous moral weight on the presence of a developmental time clock. That is, I choose to discern the presence of protectable humanity at its earliest possible moment. For me, the clock begins the instant the sperm and egg pronuclei fuse and undergo a round of replication (this takes place within twenty-four hours of the sperm and egg's first encounter). I hold to this standard because certain biochemical processes become unleashed at this time that proceed in an unalterably organized—and inexorably deterministic—fashion. This unleashing provides a substrate upon which God can transfer His worth, even if we do not know when He ultimately transfers it. It is difficult for me to think of this fact in any other way than compelling because it is impossible for me to live with consequences of its absence. I guess it doesn't bother me that this is a prejudice. Since science is untethered from this question, and not everybody is a believer, that is the realm in which we are forced to dwell.

Although I am keenly aware of my prejudices, I am also aware that molecular biology is changing the nature of how these cellular processes are perceived. Which must affect how these prejudices become draped over them. Excavating the secrets of genetic processes from individual cells will give us variations on this question that won't necessarily fit into monolithic beliefs. We must already face the ambiguities concerning genetic screening and direct interference in human chromosomal structure. In the future we will have to address the issues of conception in ways that right now are the steaming wreckage of paranoid imaginations. Which means that we haven't seen anything yet. It's funny in a way. The one thing true about science is that it usually makes our heads go further than our hearts can comprehend. Since both organs are connected, our hearts get stretched, sometimes with excruciating pain. When that happens, we usually don't wish that science would go further.

We usually wish that science would go away.

Epilogue

Putting Things in Perspective

I'd like to conclude this book by relating a tale I have heard many, many times. I have heard it told with such variation that I sense many of the details have less a foothold in fact than they do in urban legends. Nonetheless, it appears to have had at least a grain of truth thrown into it. And if we can discern the message through a bit of timeworn static, it is worth the retelling here.

The story takes place in the jungles of central Africa many years ago. This forbidding ecological sauna has incubated some very interesting inhabitants through the years. Among its most unusual citizens is a human population, the African Pygmy. This population of humans may be missing a particular human growth hormone or possibly its receptor. As a result, they are quite small when compared to Western adults.

The thick tapestries of moss, vines, and tree branches of the pygmies' jungle home are so close that they can be brushed with their small faces. The treetop canopy above them is extremely dense; sunlight is always filtered down in illuminated waterfalls of yellow-green light. These people have never seen large patches of open blue sky, complete with their great herds of African clouds. Anthropologists have studied the Pygmy and have wondered how such closed quarters affect their view of the world at large.

Years ago, with clearance from the local government, some scientists decided to find out. A group of biologists and anthropologists escorted one such tribe on a little field trip. The destination was kept secret. The only clue given was that they would be traveling to a spot the Pygmies had never visited before. Since the trip was made on foot, the journey took many days of marching through extremely dense jungle undergrowth. Eventually, one of the Caucasian scouts ahead signaled that they had reached their destination. The scientists in the troop quickly scrambled ahead to inspect.

The mystery of their destination was soon revealed. When the rest of the party caught up with the scientists, the jungle suddenly cleared. It was a dangerous clearing; the next step would cause any member of the troop to fall over the edge of a cliff almost four hundred feet straight down. They had been slogging through a jungle on top of a giant plateau and the group had reached its edge. Before them lay the incredible expanse of the vast African savannah. In the distance, tiny wildebeests could be seen grazing near a waterhole. Giant birds circled overhead. The world was bathed not in emerald green but in the sandy yellows of the open grassland.

The anthropologists, knowing that the Pygmies had never encountered such open spaces before, had run ahead to take up positions for observation. And that perhaps was fortunate because the first thing a number of the Pygmies did was to walk out onto the air over the edge of the cliff. The scientists had to catch them so that they would not fall to their deaths. The Pygmies, thus constrained at the lip of the plateau, reached their hands into the air. They curled the thumb and index finger and picked at the atmosphere like one might pick blueberries. Curiously, the motion of their hands seemed to agitate them. This picking motion was more akin to food gathering than exploration of unfamiliarity. All the observers agreed that the tribe's behavior was highly unusual. As is typical when one tries to understand foreign cultures in Western terms, the scientists were at a loss to explain the behavior.

Why did the Pygmies react to the cliff as they did? The explanation turned out to be simple. They did not attempt to walk out onto thin air because they had a death wish. Instead, they had no perception of what "down" meant in a large open space. The Pygmies had never encountered such a giant expanse before. They were unaware that the grasses they could see on the plain from the edge of the cliff were really miles away from them.

The reason for the strange picking motion turned out to come from a similar misperception. They were attempting to pick up the animals they saw in the distance and either eat them or take them home. They again had no idea of distance perception. They thought that the animals in the distance were as physically small as these animals appeared to their eyes. In the jungle they had no point of reference for such distance because every animal in their experience had always been reasonably close to them. The reason for the agitation was that they could not grasp something so obviously present.

The most interesting reaction of the day came when everything was explained to them. When they began to realize the magnitude of what they beheld on the cliff, they panicked. Terrified of the newfound perspective, they ran helter-skelter back into the safety of the jungle. It took several hours for the rest of the party to catch up to them.

Being wary of unfamiliar territory is a classic defense mechanism. So many things in our environment can harm us that it's often prudent to assume the worst until proved otherwise. I tend to be just like that, often fearing the ghosts of midnight even on sunny afternoons. And when that unfamiliarity shows real signs of being threatening, my first instinct is to run away from it as fast as I can. Thus, I have never really wondered which character I would identify with most closely in this story. I have wondered only how far I would outrun the other members of the tribe.

The same kind of unfamiliar territory exists in attempting to understand the difference between human beings and human tissues. It ends up being a strange journey into a land where the borders of the objective melt into the territory of opinion. For me, the tug-of-war about human lives and human tissues is as much an argument about beliefs as it is a discussion about facts. What is intriguing is that they can be in such opposition to each other. Oceans of scientific capabilities get tossed onto the jagged rocks of our value systems with such force that we wonder if our values will become eroded before the waves decide to break. That can be intimidating even for stout hearts. Many would rather not risk getting into the science and forging ahead into ambiguity than risk getting into the science and forging ahead into sin.

I believe this is a legitimate fear. It can be successfully argued that just because some scientific feat can be done does not automatically mean that it should be done. Indeed, a primary reason for discussing the difference between human tissue and human life centers on trying to find the difference between the irresponsible and the invaluable. If a group of human cells is a human being, many feel we have to treat it with all the respect due our own offspring and kin. If a group of cells is not a human being, working with such tissue will have all the moral ambiguity of eating a graham cracker.

I have attempted in this book to come to the edge of the ideas separating human life and human tissue. I have sought to address whether science is capable of discerning the difference between the two. Since it would be impossible to prevent, I have also tried to assess the influences of my opinions by dealing with them directly. This has meant describing how the observations I see in science conflict—or at least stretch—what I believe about God and the complexities of His creation. In motivation, I have sincerely wanted to come to a conclusion. In essence, I have simply defined my dilemmas. I am at once propelled by the challenge of science and scared at the magnitude of its implications.

If it turns out that protectable humanity begins way after conception, we can work with all kinds of tissue, creating the potential to reap more of the great benefits of research medicine. If we become guilty of experimenting on individual personalities, however, we will be responsible for some of the greatest, most hideous crimes human minds can execute.

Made all the worse because they would be done with such good intentions.

When I see the balance of these two forces suspended over the heads of what may or may not be human beings, I get queasy. All of a sudden, I find myself picking up animals that are too distant to touch. Or I begin to walk over the edge of cliffs that have no visible means of support. And then my feet itch. It makes you wonder if the Pygmies didn't have the right idea after all.